CAMBRIDGE LIBRARY COLLECTION

Books of enduring scholarly value

Zoology

Until the nineteenth century, the investigation of natural phenomena, plants and animals was considered either the preserve of elite scholars or a pastime for the leisured upper classes. As increasing academic rigour and systematisation was brought to the study of 'natural history', its subdisciplines were adopted into university curricula, and learned societies (such as the London Zoological Society, founded in 1826) were established to support research in these areas. These developments are reflected in the books reissued in this series, which describe the anatomy and characteristics of animals ranging from invertebrates to polar bears, fish to birds, in habitats from Arctic North America to the tropical forests of Malaysia. By the middle of the nineteenth century, this work and developments in research on fossils had resulted in the formulation of the theory of evolution.

A Naturalist's Calendar

The naturalist Gilbert White (1720–93) was known for his meticulous observations of flora and fauna in their natural environment, primarily around his village of Selborne in Hampshire. This posthumous 1795 publication, edited by the physician and writer John Aikin (1747–1822), comprises a collection of extracts from White's previously unpublished papers from 1768 to his death. Presented here for 'lovers of natural knowledge' is a full year of White's observations. Following the month-by-month record of natural events, the book contains brief studies of birds, quadrupeds, insects, plants and the weather. A lifelong lover of the outdoors, White had kept a near daily record of his activities for more than forty years. Regarded as one of the fathers of ecology, inspiring others to appreciate the natural world, White is best known for *The Natural History and Antiquities of Selborne* (1789), which is also reissued in the Cambridge Library Collection.

A Naturalist's Calendar

With Observations in Various Branches of Natural History

GILBERT WHITE
EDITED BY JOHN AIKIN

CAMBRIDGE
UNIVERSITY PRESS

CAMBRIDGE
UNIVERSITY PRESS

University Printing House, Cambridge, CB2 8BS, United Kingdom

Cambridge University Press is part of the University of Cambridge.
It furthers the University's mission by disseminating knowledge in the pursuit of
education, learning and research at the highest international levels of excellence.

www.cambridge.org
Information on this title: www.cambridge.org/9781108076555

© in this compilation Cambridge University Press 2014

This edition first published 1795
This digitally printed version 2014

ISBN 978-1-108-07655-5 Paperback

This book reproduces the text of the original edition. The content and language reflect
the beliefs, practices and terminology of their time, and have not been updated.

Cambridge University Press wishes to make clear that the book, unless originally published
by Cambridge, is not being republished by, in association or collaboration with,
or with the endorsement or approval of, the original publisher or its successors in title.

The original edition of this book contains a number of colour plates,
which have been reproduced in black and white. Colour versions of these
images can be found online at www.cambridge.org/9781108076555

Tnner pinx.

London: Published April 14. 1795.

J.F. Miller sculp.

A HYBRID BIRD.

A

NATURALIST'S CALENDAR,

WITH

OBSERVATIONS IN VARIOUS BRANCHES

OF

NATURAL HISTORY;

EXTRACTED FROM THE PAPERS

OF THE LATE

Rev. GILBERT WHITE, M. A.

OF SELBORNE, HAMPSHIRE,

SENIOR FELLOW OF ORIEL COLLEGE, OXFORD.

———

NEVER BEFORE PUBLISHED.

———

LONDON:

PRINTED FOR B. AND J. WHITE, HORACE'S HEAD,
FLEET STREET.

———

1795.

ADVERTISEMENT.

THE *Reverend Mr. White*, fo agreeably
known to the public by his *Natural Hiftory
of Selborne*, left behind him a feries of yearly
books, containing his diurnal obfervations
on the occurrences in the various walks of
rural nature, from the year 1768 to the time
of his death in 1793. From thefe annals he
had already extracted all the matter com-
prized in the work above mentioned, down
to the middle of 1787; but feveral curious
facts in the preceding numbers had not been
thus employed; and all the fubfequent ones
remained untouched. It was thought a mark
of refpect due to his memory, and to the re-
putation he had acquired as a faithful and
elegant obferver, not to confign thefe relicks

to neglect. The manuscripts were accordingly put into my hands for the purpose of selecting from them what might seem worthy of laying before the public. The present small publication is the fruit of my research. With no small pains I collected the materials of it, dispersed through the records of so many years, and gave them such an arrangement as I thought would present them in the most agreeable and useful manner to the lovers of natural knowledge.

J. AIKIN.

London, Jan. 1, 1795.

THE

NATURALIST'S CALENDAR.

THE
NATURALIST'S CALENDAR.

THE mode in which the following rural
calendar of the year has been compofed, was
to copy out from the journals all the circumftances thought worthy of noting, with
the feveral dates of their recurrence, and to
preferve the earlieft and lateft of thefe dates;
fo that the calendar exhibits the extreme
range of variation in the firft occurrence of
all the phenomena mentioned. To many
of them only one date is annexed, only one
obfervation having been entered. This is
particularly the cafe with refpect to the
flowering of plants, with which the book of
1768 alone was copioufly filled; and it is to
be noted that this was rather a backward year.

Of

Of the abbreviations ufed, *fl.* fignifies *flowering*; *l. leafing*; and *ap.* the firft *appearance.*

Jan.	1—12.	Redbreaft whiftles
	1—18.	Larks congregate
	1—14.	Nuthatch chatters
	1. Feb. 18.	Winter aconite fl.
	2.	Shellefs fnails ap.
	2—11.	Grey and white wag-tails ap.
	2—14.	Miffel thrufh fings
	2. Feb. 14.	Helleborus fœtidus (bears foot) fl.
	2. April 12.	Polyanthus fl.
	2. Feb. 1.	Double daify fl.
	3. Feb. 16.	Mezereon fl.
	3.	Viola tricolor (panfie) fl.
	3—21.	Lamium rubrum (red dead-nettle) fl.

Jan.

Jan. 3—15. Senecio vulg. (ground-
fel) fl.

3. Feb. 28. Hazel catkins open

4. Feb. 18. Hepatica fl.

5—12. Hedge fparrow whiftles

5. Feb. 3. Flies in windows

6. Feb. 6. Large titmoufe makes its
fpring note

6—22. Song thrufh or throftle
fings

6. Infects fwarm under funny
hedges

6. April 7. Primula vulg. (primrofe)
fl.

6. Mar. 19. Bees come out of their
hives

6. Feb. 3. Gnats play about

6—11. Hen chaffinches flock

8. Feb. 1. Ulex Europ. (gorfe) fl.

Jan.

Jan. 8. April 1: Chieranthus chieri (wall
                    flower) fl.

      8—12.   Stocks fl.

      9.   Emberiza alba (bunting)
                    in great flocks

      9.   Linnets congregate in
                    vaft flocks

      9—11.   Lambs begin to fall

      10. Feb. 21. Rooks refort to their neft
                    trees

      10.   Helleborus nigr. fl.

      10. Feb. 5. Galanthis nival. (fnow-
                    drop) fl.

      13.   Lamium alb. (white dead-
                    nettle) fl.

      13.   Trumpet honeyfuckle fl.

      13.   Ranunculus   repens
                    (creeping crowfoot) fl.

      14.   Houfe fparrows chirp

                                    Jan.

Jan. 16. Mar. 11. Leontodon   taraxacum
(dandelion) fl.

16. Mar. 24. Bat appears

16.   Spiders fhoot their webs

16.   Butterfly ap.

16.   Brambling ap.

17.   Blackbird whiftles

17.   Wren fings

18. Feb. 8. Earth worms lie out

13. Mar. 18. Crocufes fl.

21.   Skylark fings

22.   Ivy cafts its leaves

22 — 24.   Helleborus hyemalis fl.

23.   Scarabeus   ftercorarius
(common   dor or
clock) ap.

23.   Peziza acetabulum ap.

23. Mar. 5. Helleborus virid. fl.

23. Feb. 1. Hazels fhow their female
bloffoms

Jan.

Jan. 24. Feb. 21. Woodlark fings

24. Feb. 15. Chaffinch fings

25. Mar. 4. Jackdaws begin to come
to churches

25. April 14. Yellow wagtail ap.

25.   Lonicera Periclymenum
(honeyfuckle) l.

27. Mar. 15. Veronica agreftis fl.

27. April 2. Papilio urticæ (fmall tor-
toife-fhell   butterfly)
ap.

28.   White wagtail fings

28. Feb. 24. Shell-fnails ap.

30.   Earthworms engender

Feb. 1. Mar. 26. Fragaria fterilis (barren
ftrawberry) fl.

1.   Parus cæruleus (tomtit)
makes its fpring note

2.   Brown wood owls hoot

3.   Hens fit.

Feb.

Feb. 3.   Marſh titmouſe begins his two harſh ſharp notes

4. April 1. Goſſamer floats

4. April 8. Muſca tenax ap.

5.   Lauruſtinus fl.

5.   Ruſcus aculeatus (butchers broom) fl.

7.   Foxes ſmell rank

10.   Turkey cocks ſtrut and gobble

12.   Yellow hammer ſings

13. April 2. Papilio rhamni (brimſtone butterfly) ap.

13. Mar. 23. Green woodpecker laughs

14—17.   Ravens build

14. Mar. 27. Taxus baccata (male yew tree) ſheds its farina

15. Mar. 23. Tuſſilago farfara (coltsfoot) fl.

Feb.

Feb. 16. Mar. 6.   Rooks build

17.   Partridges pair

17. Mar. 8.   Peafe fown

18.   Houfe pigeons build

20. Mar. 30. Field crickets open their
                holes

21—26.   Pulex irritans (common
                flea) ap.

21. April 13. Ficaria verna (pilewort) fl.

21. April 5.   Goldfinch fings

22. Mar. 26. Coluber berus (viper) ap.

23. April 1.   Onifcus afellus (wood-
                loufe) ap.

24.   Miffel thrufhes pair

24. April 7.   Narciffus pfeudo narciffus
                (daffodil) fl.

24. April 2.   Willow fl.

25.   Frogs croak

26. Mar. 31. Viola odorata (fweet vio-
                let) fl.

                                Feb.

Feb. 26.   Phalæna tinea veſtianella ap.

27. April 24. Charadrius œdicnemus (ſtone curlew) clamours

27.   Filbert fl.

27. April 5. Ringdove cooes

28. April 17. Apricot-tree fl.

28. Mar. 24. Toad ap.

28. Mar. 22. Frogs ſpawn

Mar. 1. April 2. Veronica hederifol. fl.

2. April 17. Peach-tree fl.

2. April 6. Rana temporar. (frog) ap.

3.   Thlaſpi burſa paſt. (ſhepherds purſe) fl.

3—29. Pheaſant crows

4. May 8. Land tortoiſe comes forth

4. April 16. Pulmonaria officin. (lungwort) fl.

Mar.

Mar. 4.   Podura fimetaria ap.

4.   Aranea fcenica faliens ap.

5—16   Scolopendra forficata ap.

5. April 25.   Wryneck returns

5.   Goofe lays

5.   Duck lays

6. April 18.   Viola canina (dog vio-
                 let) fl.

6.   Papilio io (peacock but-
                 terfly) ap.

7—14.   Trouts begin to rife

8.   Beans planted

8.   Blood-worms appear in
                 the water

10.   Crows build

10—18.   Oats fown

12. April 30.   Golden-crefted wren fings

12.   Populus tremula(afpen)fl.

13—20.   Sambucus nigra (com-
                 mon elder) l.

                                    Mar.

Mar. 15. May 21. Laurel fl.

15. Chryfomela Gotting. ap.

15. April 22. Black ants ap.

16. Ephemeræ bifetæ ap.

17. April 11. Ribes Groffularia
(goofeberry) l.

17. May 19. Stellaria Holoftea (ftitch-
wort) fl.

17. April 22. Anemone nemorofa
(wood anemone) fl.

17. Blackbird fits

17. Raven fits

18—30. Wheat-ear returns

18. April 13. Adoxa mofchatellina fl.

19. April 13. Small uncrefted wren ap.

19. Fumaria bulbofa fl.

19. April 4. Ulmus campeftris (elm)
fl.

19. April 7. Turkey lays

20. Houfe pigeons fit.

B   Mar.

Mar. 20. April 14. Caltha paluſtr. (Marſh
Marigold) fl.

21. April 28. Bombylius medius ap.

21. April 12. Sand martin ap.

22.—30.   Coluber natrix (ſnake)
ap.

22. April 18. Formica Herculanea
(horſe ant) ap.

22. April 22. Green-finch ſings

23. April 14. Ivy berries ripe

25.   Vinca minor (periwinkle)
fl.

25. April 1.   Daphne Laureola (ſpurge
laurel) fl.

26. April 20. Swallows ap.

26. May 4.   Blackcap whiſtles

27.   Ducks hatched

27. April 9. Chryſoſplenium oppoſi-
tifol. fl.

28. May 1.   Houſe Martin ap.

Mar.

Mar. 28. April 13. Chimney fwallow ap.

29. April 22. Double hyacinths blow

29. Young goflings.

30. April 22. Oxalis Acetofella (wood forrel) fl.

30. April 17. Ring Ouzel ap.

31. April 20. Barley fown

April 1. May 1. Nightingale fings

1. May 4. Fraxinus excels. (Afh) fl.

1. Spider's webs on the fur-face of the ground

2.—24. Fritillaria Meleagris fl.

2. Julus terreftr. ap.

3.—24. Primula veris (Cowflip) fl.

3.—15. Glecoma hederacea (Ground ivy) fl.

3. Snipe pipes

3. Buxus (Box-tree) fl.

3. Ulmus campeft. (Elm) l.

April 3.—14.   Goofeberry fl.

3.—5.   Currant fl.

3. May 21. Pear-tree fl.

4.   Lacerta vulg. (Newt or
        Eft) ap.

5.—19.   Mercurialis   perenn.
        (Dog's Mercury) fl.

5.   Ulmus glabra fl.

6.—20.   Cardamine pratenfis
        (ladies fmock) fl.

7.—26.   Cuckow heard

7. May 10. Prunus fpinofa (Black
        thorn) fl.

7.   Termes pulfatorius beats

7.   Gudgeon fpawns

8.—28.   Ruticilla (red ftart) ap.

8.—24.   Fritillaria imper. (crown
        imperial) fl.

9.—19.   Tit lark fings

10. May 8. Fagus fylv. (beech-tree) l.

April 11. May 9. Shell fnails come out in
troops

11. Middle yellow wren ap.

13. May 7. Swift ap.

14. May 17. Conops calcitrans (ftinging fly) ap.

14. Draba verna (whitlow
grafs) fl.

14. Larch l.

14. May 14. White throat ap.

14. Red ant ap.

14. May 27. Gryllus Gryllotalpa
(Mole cricket) *churs*.

14—19—23. Second willow or laugh-
ing wren ap.

15—19. Pedicularis fylv. (red
rattle) fl.

15. Mufca carnaria (common
flefh fly) ap.

B 3 April

April 16.   Coccinella bipunctata
(lady cow) ap.

16—30.   Alauda Locuftæ vocæ
(grafshopper lark) ap.

17. May 7.   Large fhivering willow
wren ap.

17—27.   Middle willow wren (re-
gulus non criftatus me-
dius) ap.

18. May 12. Prunus Cerafus (wild
cherry) fl.

18. May 11. Garden cherry fl.

18. May 5. Prunus domeft. (plum) fl.

19—25.   Hyacinthus non fcriptus
(harebell) fl.

20—17.   Turtle cooes

20. June 11. Cratægus oxyacanthus
(hawthorn or May) fl.

21.   Orchis mafcula fl.

April

April 21. May 23. Mufca vomitoria (Blue flefh fly) ap.

22. Black fnails abound.

22. May 25. Apple-trees fl.

22. June 11. Large bat ap.

23—29. Fragaria vefca (Strawberry) fl.

23. Eryfimum Alliaria (Sauce alone) fl.

24. Prunus avium (bird cherry) fl.

24. Apis hypnorum ap.

24. May 28. Mufca meridiana ap.

25. Afilus (Wolf fly) ap.

28. May 20. Papilio Braffica (great white cabbage butterfly) ap.

30. May 21. Libellulæ (Dragon flies) ap.

30. June 6. Acer majus (fycamore) fl.

B 4 May

May 1.   Bombylius minor ap.

1. June 11. Glow-worm fhines

1—26.   Caprimulgus (fern owl or
               goat fucker) returns

1.   Ajuga reptans (bugle) fl.

2—24.   Gryllus campeft. (field
               crickets) *crink*

2—26.   Scarabæus Melolontha
               (May chaffer) ap.

3—30.   Lonicera periclymen.
               (honeyfuckle) fl.

4—12.   Lathræa fquammaria
               (toothwort) fl.

4. June 17. Shell fnails copulate

4.   Small reed fparrow fings

5—17.   Viburnum Lantana
               (mealy tree) fl.

10 – 30.   Stoparola (fly catcher) ap.

10. June 9. Apis longicornis ap.

May

May 11—13.	Paſſer Arund. minor (ſedge bird) ap.
13—15.	Oak in male bloom
13.	Papilio Atalanta (admiral butterfly) ap.
14.	Papilio cardamines (orange tip butterfly) ap.
15—26.	Fagus ſylvat. (beech) fl.
16.	Acer campeſt. (Maple) fl.
17—26.	Berberis vulg. (Barberry) fl.
17.	Papilio Ægeria (wood Argus butterfly) ap.
18. June 11.	Orange lily fl.
18. June 13.	Sphynx Filipendulæ (burnet moth) ap.
18.	Juglans regia (walnut) l.
18. June 5.	Cytiſus laburnum fl.

May

May 18. June 9. Hippobofca equina
(foreft fly) ap.

   19. June 8. Hedyfarum Onobrychis
(faintfoin) fl.

   20. June 15. Pæonia offic. (piony) fl.

   21. June 9. Æfculus Hippocaftanum
(horfe chefnut) fl.

   21. Lilac fl.

   21—27. Aquilegia vulg. (colum-
bine) fl.

   21. June 20. Mefpilus German. (med-
lar) fl.

   21. Tormentilla reptans fl.

   22. Convallaria min. (lily
of the valley) fl.

   22. July 22. Bees fwarm

   22—25. Afperula odorat. (wood
roofe) fl.

   23. Wafps (female) ap.

May

May 23. June 8. Sorbus aucuparia (mountain afh) fl.

24. June 11. Ophrys nidus avis (birds neft orchis) fl.

24. June 4. Cratægus Aria (white beam tree) fl.

24. June 7. Polygala vulg. (milkwort) fl.

25. Ciftus Helianthemum (rock-rofe) fl.

26. Viburnum Opulus (Gelder rofe) fl.

26. June 25. Sambucus niger (elder) fl.

26. Cantharis noctiluca ap.

27. June 9. Apis longicornis bores holes in walks.

27. June 13. Morus nigra (mulberry) l.

27. Cratægus torminalis (wild fervice tree) fl.

May

May 27. June 13. Sanicula europæa fl.

28. Geum urbanum (avens) fl.

28. Orchis morio fl.

29. June 1. Lychnis flos cuculi (cuckow-flower) fl.

29. Poterium Sanguiſorba (burnet) fl.

30. June 22. Digitalis purpur. (fox-glove) fl.

30. June 20. Corn flag fl.

30. June 13. Serapias longifol. fl.

30. June 21. Rubus Idæus (raſpberry) fl.

30. Geranium robertianum (herb robert) fl.

31. Scrophularia nodoſa (figwort) fl.

31. Lithoſpermum officin. (gromwell) fl.

June

June 1.   Euphorbia Amygdaloid.
              (wood ſpurge) fl.

     1.   Allium urſinum (ram-
              ſons) fl.

     1.   Myoſotis ſcorpoides
              (mouſe-ear ſcorpion-
              graſs) fl.

  1—14.   Graſshopper ap.

  1—21.   Roſe fl.

  1 July 16. Hieracium minor fl.

     1.   Menyanthes trifol.(buck-
              bean) fl.

  2—8.   Scarabæus aurat. (braſs
              or green beetle) ap.

  2—23.   Sheep ſhorn

     2.   Iris Pſeudacorus (water
              flag) fl.

     2.   Secale Cereale (Rye) fl.

     2.   Cynogloſſum offic.
              (hounds-tongue) fl.

                            June

June 2. Aug. 6. Serapias latifol. fl.

2.   Mufca Cæfar (green-gold
           fly) ap.

2.   Papilio Moera (Argus
           butterfly) ap.

3.   Ranunculus flammula fl.

3.   Lotus cornicul. (birds
           foot trefoil) fl.

3—11.   Fraxinella fl.

3.   Phryganea nigra ap.

3—14.   Ephemera vúlg. (angler's
           May fly) ap.

4.   Anthyllis vulner. (ladies
           fingers) fl.

4. July 4. Ophrys apifera (bee
           orchis) fl.

5—19.   Pinks fl.

5.   Philadelphus coronar. fl.

5—20.   Libellula Virgo ap.

7. July 30. Vitis vinifera (vine) fl.

June

June 8. July 1. Portugal laurel fl.

8—25. Purple Martagon fl.

8. Aug. 1. Geranium pratenſe (meadow craneſbill) fl.

8. Tamus communis (lady ſeal) fl.

9. Field pea fl.

9. Cucubalus behen. (bladder campion) fl.

9. Bryonia alba (bryony) fl.

10. Stachys ſylvat. fl.

11. Solanum Dulcamara (bitter-ſweet) fl.

12. Juglans reg. (walnut) fl.

12. July 23. Phallus impudicus ap.

12. Epilobium anguſtifol. (narrow-leaved willow-herb) fl.

13. July 22. Triticum hybern. (wheat) fl.

June

June 13.	Symphytum offic. (comfrey) fl.
13—30.	Lyſimachia nemorum fl.
15. Aug. 24.	Tremella noſtoc ap.
16.	Rhamnus cathart. (buckthorn) fl.
16.	Cicada ſpumaria (cuckow-ſpit infeɗ or froghopper) ap.
17—18.	Roſa canina (dog-roſe) fl.
17. Sept. 3.	Lycoperdon boviſta (large puff-ball) ap.
18.	Verbaſcum Thapſus (Mullein) fl.
19.	Echium vulg. (Viper's bugloſs) fl.
19. July 20.	Meadow hay cut
19.	Scarabæus cervus ap.
20.	Borago officin. fl.

June

June 20.	Euonymus europæus (Spindle tree) fl.
20. July 4.	Carduus nutans (muſk thiſtle) fl.
21.	Cornus ſanguin. (dog-wood) fl.
21.	Scabioſa arv. (field ſcabi-ous) fl.
21—27.	Carduus paluſtris (marſh thiſtle) fl.
22. July 9.	Spiræa Filipendula (drop-wort) fl.
22. July 7.	Valeriana officin. fl.
22. July 4.	Quails call
22.	Epilobium montan. (mountain willow-herb) fl.
23—29.	Carduus criſpus (thiſtle upon thiſtle) fl.

C   June

June 23.		Heracleum Sphondylium (cow parfnep) fl.
23.		Bunium Bulbocaft. (earth nut) fl.
23.	Aug. 2.	Young frogs migrate
24.		Oeftrus curvicauda ap.
24.		Verbena officin. (Vervain) fl.
24.		Papaver Rhoeas (corn poppy) fl.
24.		Prunella vulg. (felf-heal) fl.
24—29.		Agrimonia Eupator. (Agrimony) fl.
24.	Aug. 2.	Tabanus Bovinus (great horfe-fly) ap.
25.		Centaurea Scabiofa (great knapweed) fl.
26.	Aug. 30.	Agaricus campeft. (mufhroom) ap.

June

June 26.   Malva fylv. (common
                mallow) fl.

26.   —— rotundifol. fl.

26.   Hypericum perforat. (St.
                John's wort) fl.

27. July 4.   Orobanche major (broom
                rape) fl.

27.   Hyofcyamus niger (hen-
                bane) fl.

27.   Tragopogon pratens.
                (goat's beard) fl.

27.   Atropa Belladonna
                (deadly nightfhade) fl.

28. July 29.   Truffles begin to be found.

28. July 31.   Young partridges fly

28. July 31.   Tilia europ. (lime tree)
                fl.

28. July 12.   Carduus lanceolat. (fpear
                thiftle) fl.

                C 2   June

June 28.		Spiræa Ulmaria (mea-dow-fweet) fl.
28.		Genifta tinctoria (Dyers broom) fl.
28.		Thymus ferpyllum (wild thyme) fl.
29.	July 20.	Stachys germanic. fl.
29.	July 4.	Hemerocallis(daylily)fl.
29.	July 30.	Jafmine fl.
29.	Aug. 4.	Hollyhock fl.
29.	July 23.	Monotropa Hypopithys fl.
29.		Galium verum (ladies bedftraw) fl.
29.		——— paluftre fl.
29.		Lapfana com. (nipple-wort) fl.
29.		Carduus acanthoides (welted thiftle) fl.

June

June 30.   Achillea Ptarmica
(fneezewort) fl.

30.   Malva mofchat. (mufk
mallow) fl.

30.   Anagallis arv. (pimper-
nel) fl.

30. July 17. Scarabæus folftit. (hoary
beetle) ap.

July 1.   Serratula arv. (common
thiftle) fl.

1.   Adonis ann. (pheafant's
eye) fl.

2.   Euphrafia Odontit. (red
eyebright) fl.

2.   Bupleurum rotundifol.
(thorough-wax) fl.

2.   Agroftemma Githago
(cockle) fl.

2.   Prenanthes muralis (ivy-
leaf) fl.

C 3   July

July	2.	Matricaria Parthenium (feverfew) fl.
	3.	Sedum acre (ftone crop) fl.
	3.	Liguftrum vulg. (Privet) fl.
	3.	Antirrhinum Linaria (toad flax) fl.
	4.	Linum perenne (fiberian flax) fl.
	4—24.	Vaccinium ulig. (whortle-berries) ripe
	5.	Refeda lutea fl.
	5.	Centaurea Cyanus (blue bottle) fl.
	5—12.	Carduus acaulis (dwarf carline thiftle) fl.
	6.	Typha latifol. (bulrufh) fl.
	6.	Lythrum Salicaria (fpiked willow-herb) fl.

July

July 6.   Verbaſcum niger. (black
                   mullein) fl.

      6.   Chryſanthemum fl.

      6—9.   Marigolds fl.

      7.   Sherardia arv. (little field
                   madder) fl.

      7.   Meliſſa Nepeta (field ca-
                   lamint) fl.

      7.   Balotta nigr. (henbit) fl.

      8—19.   Betonica officin. (betony)
                   fl.

      8.   Campanula rotundifol. fl.

      8.   Chenopodium Bonus
                   Henricus (Engliſh
                   mercury) fl.

      8.   Daucus carota (wild
                   carrot) fl.

      8—20.   Tropeolum maj. (Indian
                   creſs) fl.

                   C 4   July

July 9.   Nepeta cataria (cat mint) fl.

9.   Melampyrum fylvat. (cow wheat) fl.

9.   Valantia cruciata (croff-wort) fl.

9—27.   Cranberries ripe

10.   Vicia Cracca (tufted vetch) fl.

10.   —— Sylvat. (wood vetch) fl.

11.   Campanula glomerata (little throatwort) fl.

11.   Jafione montan. (hairy fheep's fcabious) fl.

12.   Paftinaca fylv. fl.

12.   Lilium alb. (white lily) fl.

13.   Conium maculat. (hem-lock) fl.

July

July 13.   Caucalis Anthrifcus fl.

13. Aug. 11. Flying ants ap.

13.   Lyfimachia Nummularia
            (moneywort) fl.

14. Aug. 4. Scarlet martagon fl.

14.   Stellaria graminea fl.

14.   Æthufa Cynapium (fool's
            parfley) fl.

14—29.   Sambucus Ebulus (dwarf
            elder) fl.

14. Aug. 29. Young martins and
            fwallows begin to
            congregate.

14.   Potatoes fl.

15.   Angelica fylv. fl.

15—25.   Digitalis ferrugin. fl.

15.   Senecio Jacobæa (rag-
            wort) fl.

15.   Solidago Virg-aurea
            (golden-rod) fl.

                         July

July 16. Centaurea Calcitrapa
     (ftar thiftle) fl.

16.   Oenothera biennis fl.

17. Aug. 14. Peafe cut

17.   Galega officin. fl.

17. Aug. 21. Apricots ripe

17.   Stachys paluftr. fl.

17.   Epilobium ramof.
     (branching willow-
     herb) fl.

17. Aug. 7. Rye harveft begins

18. Aug. 15. Chlora perfol. (yellow
     centaury) fl.

18.   Lathyrus Aphaca (yellow
     vetchling) fl.

18.   Circæa lutetiana (en-
     chanter's nightfhade) fl.

18.   Eupatorium cannabin.
     (water-hemp agri-
     mony) fl.

        July

July 19.	Campanula latifol. (giant throat-wort) fl.
19.	Euphrafia officin. (eye-bright) fl.
19. Aug. 10.	Humulus Lupulus (hop) fl.
19.	Poultry moult
20.	Cufcuta europ. (Dodder) fl.
20.	Gentianum centaureum (leffer centaury) fl.
20.	Sium nodiflorum fl.
21.	Spergula arv. (fpurrey) fl.
21.	Trifolium arv. fl.
21.	Polygonum Fagopyr. (buck wheat) fl.
21. Aug. 23.	Wheat harveft begins
22.	Sparganium erect. (great bur-reed) fl.

July

July 22—31.   Hypericum elodes (marſh St. John's wort) fl.

22.   Droſera rotundifol. (ſun dew) fl.

22.   Comarum paluſtr. (purple marſh cinquefoil) fl.

22.   Wild cherries ripe

22.   Anthericum Oſſifragum (Lancaſhire aſphodel) fl.

23.   Scutellaria galericulat. (hooded willow-herb) fl.

23.   Oenanthe fiſtulos. (water dropwort) fl.

23.   Marrubium vulg. (horehound) fl.

24.   Seſeli caruifol. fl.

24.   Aliſma Plantago (water plantain) fl.

25.   Alopecurus myoſuroides fl.

July

July 25. Aug. 9. Clematis Vitalba (vir-
gin's bower) fl.

25.   Bees kill the drones

26.   Dipſacus ſylv. (teaſel) fl.

26.   Origanum vulg. (wild
marjoram) fl.

27—29.   Swifts begin to depart

28—29.   Dipſacus piloſus (ſmall
wild teaſel) fl.

28.   Teucrium Scorodonia
(wood ſage) fl.

28.   Lathyrus latifol. (ever-
laſting pea) fl.

29.   Hypericum humifuſum
(trailing St. John's
wort) fl.

30.   Veratrum album. (white
hellebore) fl.

30.   Anthemis nobil. (camo-
mile) fl.

July

July 30.   Scabiofa columbaria fl.

31. Aug. 6.   Helianthus multiflor.
                    (fun flower) fl.

31.   Lyfimachia vulg. (yellow
                    loofe-ftrife) fl.

31. Aug. 27. Swifts laft feen

Aug. 1—16.   Oats cut

1—26.   Barley cut

1.   Scutellaria minor fl.

2.   Inula dyfenterica (marfh
                    flea-bane) fl.

2.   Apis manicata ap.

2.   Papilio machaon (fwallow
                    tailed butterfly) ap.

3—19.   Oeftrus Bovis (whame or
                    burrel fly) lays eggs
                    on horfes.

3.   Sonchus arvens. (fow-
                    thiftle) fl.

                                    Aug.

Aug. 3. Papilio Cinxia (plantain fritillary) ap.

4. Picris Hieracioides (yellow fuccory) fl.

4. Mufca myftacea ap.

5. Campanula trachelium (Canterbury bells) fl.

5. Mentha longifol. fl.

7. Carlina vulg. (carline thiftle) fl.

7. Rhus Cotinus fl.

7. Ptinus pectinicornis ap.

8. Arctium lappa (burdock) fl.

8. Sept. 3. Gentiana Amarella (fellwort) fl.

8. Artemifia Abfinthium (wormwood) fl.

8. ———— vulg. (mugwort) fl.

Aug.

Aug. 10.   Centaurea folftit. (St.
Barnaby's thiftle) fl.

10. Sept. 13. Colchicum autumn.
(meadow faffron) fl.

12. Sept. 27. After (michaelmas daify)
fl.

14.   Thalictrum flavum (meadow rue) fl.

14.   Eryngium marit. (fea
holly) fl.

14. Sept. 28. China-afters fl.

14.   Boletus albus ap.

15.   Campanula hybrida (lefs
Venus looking-glafs) fl.

15.   Carthamus tinctor. fl.

15.   Young broods of gold-
finches ap.

15. Sept. 12. Lapwings congregate.

15.   Papilio femele (black-
eyed marble butterfly)
ap.

4   Aug.

Aug. 16.   Birds reaffume their
            fpring notes
17.   Scabiofa fuccifa (devils-
            bit) fl.
17. Sept. 10. Thiftle down floats
18.   Conyza fquarrofa (plow-
            man's fpikenard) fl.
18.   Leontodon autumn. (au-
            tumnal dandelion) fl.
18.   Flies abound in windows
18. Nov. 1. Linnets congregate
20.   Bulls make their fhrill au-
            tumnal noife
22.   After amellus fl.
23.   Impatiens balfamina
            (balfam) fl.
24.   Carduus marianus (milk
            thiftle) fl.
24. Sept. 17. Hop picking begins
            D   Aug.

Aug. 24. Sept. 22. Beeches begin to be
tinged with yellow

25.   Saponaria officin. (foap-
wort) fl.

27. Sept. 12. Ophrys fpiralis (ladies
traces) fl.

29.   Papilio Phlæas (fmall
golden black-fpotted
butterfly) ap.

29.   Swallows fing

30. Sept. 2. Hybifcus Syriacus fl.

30.   Papilio Paphia (great
fritillary) ap.

31.   Phalæna paɗa (willow
red under-wing moth)
ap.

Sept. 1. Nov. 7. Stone curlews clamour

1.   Phalæna ruffula ap.

4. Oɗ. 24. Grapes ripen

Sept.

Sept. 4. Nov. 9. Wood owls hoot

    4. Papilio Hyale (faffron butterfly) ap.

    4—30. Ring Ouzle ap. on its autumnal vifit

    6—29. Stoparola (flycatcher) withdraws.

    11. Bean harveft begins

    12. Oct. 2. Hedera Helix (ivy) fl.

    12. Nov. 1. Stares congregate

    25. Wild honeyfuckles fl. a fecond time

    28. Oct. 24. Wood lark fings.

    29. Nov. 11. Woodcocks come

Oct. 1. Arbutus Unedo fl.

    3. Nov. 9. Wheat fown.

    4. Nov. 5. Swallows laft feen.

        (N. B. The houfe-martin, the lateft)

    10. Nov. 10. Redwings come

Oct.

Oct. 12. Nov. 23. Fieldfares come

15—27.   Goffamer fills the air

19.   China Hollyhock fl.

20. Dec. 31. Hen chaffinches congregate

23. Dec. 27. Wood pigeons come

23. Nov. 29. Grey crows come

25. Nov. 20. Snipes come up into the
meadows

27. Nov. 26. Tortoife begirts to bury
himfelf

31. Dec. 25. Rooks vifit their neft
trees

Nov. 1.   Bucks grunt

10.   Primrofe fl.

13—14.   Green whiftling plover
ap.

16.   Helvella mitra ap.

27.   Greenfinches flock

30. Dec. 29. Hepatica fl.

Dec.

Dec. 4—21.	Ulex europ. (gorse or furze) fl.
7—16.	Polyanthus fl.
11—27.	Lambs fall
12—23.	Moles work in throwing up hillocks
14—30.	Helleborus fœtid. fl.
15.	Daisy fl.
15.	Wall flower fl.
15.	Mezereon fl.
29.	Snowdrop fl.

" IN SESE VERTITUR ANNUS."

OBSERVATIONS

ON

VARIOUS PARTS OF NATURE.

D 4

OBSERVATIONS ON BIRDS.

BIRDS IN GENERAL.

In fevere weather, fieldfares, red wings, fky larks, and tit larks refort to watered meadows for food; the latter wades up to its belly in purfuit of the pupæ of infects, and runs along upon the floating grafs and weeds. Many gnats are on the fnow near the water, thefe fupport the birds in part.

Birds are much influenced in their choice of food by colour, for though white currants are a much fweeter fruit than red, yet they feldom touch the former till they have devoured every bunch of the latter.

Red-

Red-ftarts, Fly-catchers and Black-caps arrive early in April. If thefe little delicate beings are birds of paffage, (as we have rea-fon to fuppofe they are, becaufe they are never feen in winter) how could they, feeble as they feem, bear up againft fuch ftorms of fnow and rain, and make their way through fuch meteorous turbulencies, as one fhould fuppofe would embarrafs and retard the moft hardy and refolute of the winged nation? Yet they keep their ap-pointed times and feafons; and in fpite of frofts and winds return to their ftations pe-riodically as if they had met with nothing to obftruct them. The withdrawing and appearance of the *fhort winged* fummer birds is a very puzzling circumftance in natural hiftory!

When the boys bring me wafps nefts, my bantam fowls fare delicioufly, and when the

combs

combs are pulled to pieces, devour the young wafps in their maggot ftate with the higheft glee and delight. Any infect-eating bird would do the fame; and therefore I have often wondered that the accurate Mr. Ray fhould call one fpecies of buzzard *buteo apivorus* five *vefpivorus*, or the *honey buzzard*, becaufe fome combs of wafps happened to be found in one of their nefts. The combs were conveyed thither doubtlefs for the fake of the maggots or nymphs, and not for their honey: fince none is to be found in the combs of wafps. Birds of prey occafionally feed on infects: thus have I feen a tame kite picking up the female ants full of eggs, with much fatisfaction.

ROOKS.

Rooks are continually fighting and pulling each others nefts to pieces: thefe proceed-
ings

7

ings are inconfiftent with living in fuch clofe community. And yet if a pair offers to build on a fingle tree, the neft is plundered and demolifhed at once. Some rooks rooft on their neft trees. The twigs which the rooks drop in building fupply the poor with brufh-wood to light their fires. Some unhappy pairs are not permitted to finifh any neft till the reft have completed their building. As foon as they get a few fticks together, a party comes and demolifhes the whole. As foon as rooks have finifhed their nefts, and before they lay, the cocks begin to feed the hens, who receive their bounty with a fondling, tremulous voice, and fluttering wings, and all the little blandifhments that are expreffed by the young while in a helplefs ftate. This gallant deportment of the males is continued through the whole feafon of incubation. Thefe birds

do

do not copulate on trees, nor in their nefts, but on the ground in open fields.

THRUSHES.

THRUSHES during long droughts are of great fervice in hunting out fhell fnails, which they pull in pieces for their young, and are thereby very ferviceable in gardens.

Miffel thrufhes do not deftroy the fruit in gardens like the other fpecies of turdi, but feed on the berries of miffeltoe, and in the fpring on ivy berries which then begin to ripen. In the fummer, when their young become fledge, they leave neighbourhoods, and retire to fheep walks and wild commons.

The magpies, when they have young, deftroy the broods of miffel thrufhes, though the dams are fierce birds, and fight boldly in defence of their nefts. It is probably to

avoid

avoid fuch infults, that this fpecies of thrufh,
though wild at other times, delights to build
near houfes, and in frequented walks, and
gardens.

POULTRY.

MANY creatures are endowed with a ready
difcernment to fee what will turn to their
own advantage and emolument; and often
difcover more fagacity than could be ex-
pected. Thus my neighbour's poultry watch
for waggons loaded with wheat, and running
after them, pick up a number of grains
which are fhaken from the fheaves by the
agitation of the carriages. Thus, when my
brother ufed to take down his gun to fhoot
fparrows, his cats would run out before
him, to be ready to catch up the birds as
they fell.

The earneft and early propenfity of the
gallinæ

gallinæ to rooſt on high is very obſervable;
and diſcovers a ſtrong dread impreſſed on
their ſpirits reſpecting vermin that may
annoy them on the ground during the hours
of darkneſs. Hence, poultry, if left to themſelves and not houſed, will perch the winter
through on yew-trees and fir-trees; and
turkies, and guinea fowls, heavy as they are,
get up into apple-trees: pheaſants alſo in
woods ſleep on trees to avoid foxes; while
pea fowls climb to the tops of the higheſt
trees round their owner's houſe for ſecurity,
let the weather be ever ſo cold or blowing.
Partridges, it is true, rooſt on the ground, not
having the faculty of perching; but then the
ſame fear prevails in their minds; for
through apprehenſions from pole-cats and
ſtoats, they never truſt themſelves to coverts; but neſtle together in the midſt of
large fields, far removed from hedges and
coppices,

coppices, which they love to haunt in the day, and where at that feafon they can fculk more fecure from the ravages of rapacious birds.

As to ducks and geefe, their awkward, fplay, web feet forbid them to fettle on trees; they therefore, in the hours of darknefs and danger, betake themfelves to their own ele-ment, the water, where, amidft large lakes and pools, like fhips riding at anchor, they float the whole night long in peace and fe-curity.

HEN PARTRIDGE.

A HEN Partridge came out of a ditch, and ran along fhivering with her wings, and cry-ing out, as if wounded and unable to get from us. While the dam acted this dif trefs, the boy who attended me faw her brood, that was fmall and unable to fly, run

4   for

for fhelter into an old fox-earth under the bank. So wonderful a power is inftinct !

A HYBRID PHEASANT.

Lord Stawell fent me from the great lodge in the holt a curious bird for my infpection. It was found by the fpaniels of one of his keepers in a coppice, and fhot on the wing. The fhape, and air, and habit of the bird, and the fcarlet ring round the eyes, agreed well with the appearance of a cock pheafant; but then the head and neck and breaft and belly were of a gloffy black : and though it weighed 3lb. 3½ oz. * the weight of a large full grown cock pheafant, yet there were no figns of any fpurs on the legs, as is ufual with all grown cock pheafants, who have long ones. The legs and feet were naked of feathers; and therefore it could be nothing of the grous kind. In

* Hen pheafants ufually weigh only 2lb. 1oz.

E   the

the tail were no long bending feathers, fuch
as cock pheafants ufually have; and are cha-
racteriftic of the fex. The tail was much
fhorter than the tail of a hen pheafant, and
blunt and fquare at the end. The back,
wing feathers, and tail, were all of a pale
ruffet curioufly ftreaked, fomewhat like the
upper parts of a hen partridge. I returned
it, with my verdict, that it was probably a
fpurious or hybrid hen bird, bred between
a cock pheafant and fome domeftic fowl.
When I came to talk with the keeper who
brought it, he told me that fome peahens had
been known laft fummer to haunt the coppices
and coverts where this mule was found.

Mr. Elmer, of Farnham, the famous game
painter, was employed to take an exact copy
of this curious bird.

N. B. It ought to be mentioned, that
fome good judges have imagined this bird to
have been a ftray grous or black-cock; it is
however

however to be obferved, that Mr. W. remarks that its legs and feet were naked, where-
as thofe of the grous are feathered to the toes.

LAND RAIL.

A MAN brought me a *land-rail* or *daker-*
ben, a bird fo rare in this diftrict, that we
feldom fee more than one or two in a feafon, and thofe only in autumn. This is
deemed a bird of paffage by all the writers:
yet from its formation feems to be poorly
qualified for migration; for its wings are
fhort, and placed fo forward, and out of
the centre of gravity, that it flies in a very
heavy and embarraffed manner, with its legs
hanging down; and can hardly be fprung a
fecond time, as it runs very faft, and feems
to depend more on the fwiftnefs of its feet
than on its flying.

When we came to draw it, we found the

E 2   entrails

entrails fo foft and tender that in appearance
they might have been dreffed like the ropes
of a woodcock. The craw or crop was
fmall and lank, containing a mucus; the
gizzard thick and ftrong, and filled with
fmall fhell fnails, fome whole, and many
ground to pieces through the attrition which
is occafioned by the mufcular force and motion of that inteftine. We faw no gravels
among the food: perhaps the fhell fnails
might perform the functions of gravels or
pebbles, and might grind one another. *Land-
rails* ufed to abound formerly, I remember,
in the low wet bean fields of Chriftian-
Malford in North Wilts, and in the meadows near Paradife gardens at Oxford, where
I have often heard them cry crex, crex.
The bird mentioned above weighed $7\frac{1}{2}$ oz.
was fat and tender, and in flavour like the
flefh of a woodcock. The liver was very
large and delicate.

4   FOOD

FOOD OF THE RING DOVE.

ONE of my neighbours fhot a ring-dove on an evening as it was returning from feed and going to rooft. When his wife had picked and drawn it, fhe found its craw ftuffed with the moft nice and tender tops of turnips. Thefe fhe wafhed and boiled, and fo fate down to a choice and delicate plate of greens, culled and provided in this extraordinary manner.

Hence we may fee that graminivorous birds, when grain fails, can fubfift on the leaves of vegetables. There is reafon to fuppofe that they would not long be healthy without, for turkies, though corn fed, delight in a variety of plants, fuch as cabbage, lettuce, endive, &c. And poultry pick much grafs; while geefe live for months together on commons by grazing alone.

" Nought

" Nought is ufelefs made ;
    " On the barren heath
" The fhepherd tends his flock that daily crop
" Their verdant dinner from the moffy turf
" Sufficient: after them the cackling *goofe*,
" Clofe-grazer, finds wherewith to eafe her want."
                                    PHILIPS'S CYDER.

HEN HARRIER.

MR. WHITE of Newton fprung a pheafant in a wheat ftubble, and fhot at it;
when, notwithftanding the report of the gun,
it was immediately purfued by the blue
hawk, known by the name of the *hen-
harrier*, but efcaped into fome covert. He
then fprung a fecond, and a third, in the fame
field, that got away in the fame manner;
the hawk hovering round him all the while
that he was beating the field, confcious no
doubt of the game that lurked in the
ftubble. Hence we may conclude that this
bird of prey was rendered very daring and
                                            bold

bold by hunger, and that hawks cannot always feize their game when they pleafe. We may farther obferve, that they cannot pounce their quarry on the ground where it might be able to make a ftout refiftance, fince fo large a fowl as a pheafant could not but be vifible to the piercing eye of a hawk when hovering over a field. Hence that propenfity of cowring and fquatting till they are almoft trod on, which no doubt was intended as a mode of fecurity: though long rendered deftructive to the whole race of gallinæ by the invention of nets and guns,

GREAT SPECKLED DIVER, or LOON.

As one of my neighbours was traverfing Wolmer foreft from Bramfhot acrofs the moors, he found a large uncommon bird fluttering in the heath, but not wounded,

E 4 which

which he brought home alive. On examination it proved to be *Colymbus glacialis Linn:* the great fpeckled diver or loon, which is moſt excellently defcribed in Willughby's ornithology.

Every part and proportion of this bird is fo incomparably adapted to its mode of life, that in no inſtance do we fee the wifdom of God in the creation to more advantage. The head is ſharp, and fmaller than the part of the neck adjoining, in order that it may pierce the water; the wings are placed forward and out of the center of gravity, for a purpofe which ſhall be noticed hereafter; the thighs quite at the podex, in order to facilitate diving; and the legs are flat, and as ſharp backwards almoſt as the edge of a knife, that in ſtriking they may eafily cut the water; while the feet are palmated, and broad for fwimming, yet fo

folded

folded up when advanced forward to take a frefh ftroke, as to be full as narrow as the fhank. The two exterior toes of the feet are longeft; the nails flat and broad refembling the human, which give ftrength and increafe the power of fwimming. The foot, when expanded, is not at right angles to the leg or body of the bird: but the exterior part inclining towards the head forms an acute angle with the body; the intention being not to give motion in the line of the legs themfelves, but by the combined impulfe of both in an intermediate line, the line of the body.

Moft people know, that have obferved at all, that the fwimming of birds is nothing more than a walking in the water, where one foot fucceeds the other as on the land, yet no one, as far as I am aware, has re-

marked

marked that diving fowls, while under water, impel and row themfelves forward by a motion of their wings, as well as by the impulfe of their feet: but fuch is really the cafe, as any perfon may eafily be convinced who will obferve ducks when hunted by dogs in a clear pond. Nor do I know that any one has given a reafon why the wings of diving fowls are placed fo forward: doubtlefs, not for the purpofe of promoting their fpeed in flying, fince that pofition certainly impedes it; but probably for the increafe of their motion under water, by the ufe of four oars inftead of two; yet were the wings and feet nearer together, as in land birds, they would, when in action, rather hinder than affift one another.

This *Colymbus* was of confiderable bulk, weighing only three drachms fhort of three

6   pounds

pounds avoirdupois. It meafured in length from the bill to the tail (which was very fhort) two feet; and to the extremities of the toes four inches more; and the breadth of the wings expanded was 42 inches. A perfon attempted to eat the body, but found it very ftrong and rancid, as is the flefh of all birds living on fifh. *Divers* or *Loons*, though bred in the moft northerly parts of Europe, yet are feen with us in very fevere winters; and on the Thames are called *fprat loons*, becaufe they prey much on that fort of fifh.

The legs of the *Colymbi* and *Mergi* are placed fo very backward, and fo out of all center of gravity, that thefe birds cannot walk at all. They are called by Linnæus *compedes*, becaufe they move on the ground as if fhackled, or fettered.

STONE

STONE CURLEW.

On the 27th of February 1788, *Stone Curlews* were heard to pipe; and on March 1ft, after it was dark, fome were paffing over the village, as might be perceived by their quick fhort note, which they ufe in their nocturnal excurfions by way of watch-word, that they may not ftray and lofe their companions.

Thus, we fee, that retire whitherfoever they may in the winter, they return again early in the fpring, and are, as it now appears, the firft fummer birds that come back. Perhaps the mildnefs of the feafon may have quickened the emigration of the curlews this year.

They fpend the day in high elevated fields and fheep-walks; but feem to defcend in the night to ftreams and meadows, perhaps for

water,

water, which their upland haunts do not afford them.

THE SMALLEST WILLOW WREN.

THE fmalleft uncrefted or willow wren, or chiff chaf, is the next early fummer bird which we have remarked; it utters two fharp piercing notes, fo loud in hollow woods as to occafion an echo, and is ufually firft heard about the 20th of March.

FERN OWL, OR GOAT SUCKER.

THE country people have a notion that the *fern owl*, or *churn owl*, or *eve-jarr*, which they alfo call a *puckeridge*, is very injurious to weanling calves, by inflicting, as it ftrikes at them, the fatal diftemper known to cow-leeches by the name of *puckeridge*. Thus does this harmlefs, ill-fated bird fall under a double imputation which it by no means deferves;

deserves; in Italy, of sucking the teats of
goats, whence it is called *caprimulgus*; and
with us, of communicating a deadly diforder
to cattle. But the truth of the matter is, the
malady above mentioned is occafioned by
the *Œftrus bovis*, a dipterous infect which
lays its eggs along the chines of kine, where
the maggots, when hatched, eat their way
through the hide of the beaft into the flefh,
and grow to a very large fize. I have juft
talked with a man, who fays, he has more
than once ftripped calves who have died of
the *puckeridge*; that the ail or complaint lay
along the chine, where the flefh was much
fwelled, and filled with purulent matter.
Once I myfelf faw a large rough maggot of
this fort fqueezed out of the back of a cow.

These maggots in Effex are called wornils.

The leaft obfervation and attention would
convince men, that thefe birds neither injure

the

the goatherd nor the grazier, but are perfectly harmless, and subsist alone, being night
birds, on night insects, such as *scarabæi* and
phalænæ; and through the month of July
mostly on the *scarabæus solstitialis*, which in
many districts abounds at that season. Those
that we have opened, have always had their
craws stuffed with large night-moths and
their eggs, and pieces of chaffers: nor does
it anywise appear how they can, weak and
unarmed as they seem, inflict any harm
upon kine, unless they possess the powers of
animal magnetism, and can affect them by
fluttering over them.

A *fern owl*, this evening (August 27),
showed off in a very unusual and entertaining
manner, by hawking round and round the
circumference of my great spreading oak,
for twenty times following, keeping mostly
close to the grass, but occasionally glanc-
ing

ing up amidft the boughs of the tree. This amufing bird was then in purfuit of a brood of fome particular phalæna belonging to the oak, of which there are feveral forts; and exhibited on the occafion a command of wing fuperior, I think, to that of the fwallow itfelf.

When a perfon approaches the haunt of fern-owls in an evening, they continue fly-ing round the head of the obtruder, and by ftriking their wings together above their backs, in the manner that the pigeons called fmiters are known to do, make a fmart fnap: perhaps at that time they are jealous for their young; and this noife and gefture are intended by way of menace.

Fern-owls have attachment to oaks, no doubt on account of food; for the next evening we faw one again feveral times among the boughs of the fame tree; but it

did

did not fkim round its ftem over the grafs, as on the evening before. In May thefe birds find the *Scarabæus melolontha* on the oak; and the *Scarabæus folftitialis* at Mid-fummer. Thefe peculiar-birds can only be watched and obferved for two hours in the twenty-four; and then in a dubious twilight, an hour after fun-fet and an hour before fun-rife.

On this day (July 14, 1789) a woman brought me two eggs of a *fern-owl*, or *eve-jarr*, which fhe found on the verge of the hanger, to the left of the hermitage under a beechen fhrub. This perfon, who lives juft at the foot of the hanger, feems well acquainted with thefe nocturnal fwallows, and fays fhe has often found their eggs near that place, and that they lay only two at a time on the bare ground. The eggs were oblong, dufky, and ftreaked fomewhat in the

F   manner

manner of the plumage of the parent bird, and were equal in fize at each end. The dam was fitting on the eggs when found, which contained the rudiments of young, and would have been hatched perhaps in a week. From hence we may fee the time of their breeding, which correfponds pretty well with that of the fwift, as does alfo the period of their arrival. Each fpecies is ufually feen about the beginning of May. Each breeds but once in a fummer; each lays only two eggs.

July 4, 1790. The woman who brought me two *Fern-owls* eggs laft year on July 14, on this day produced me two more, one of which had been laid this morning, as appears plainly, becaufe there was only one in the neft the evening before. They were found, as laft July, on the verge of the down above the hermitage under a beechen fhrub,

on

on the naked ground.—Laſt year thoſe eggs were full of young, and juſt ready to be hatched.

Theſe circumſtances point out the exact time when theſe curious nocturnal migratory birds lay their eggs, and hatch their young.

Fern·owls, like ſnipes, ſtone curlews, and ſome other birds, make no neſt. Birds that build on the ground do not make much of neſts.

SAND MARTINS.

MARCH 23, 1788. A gentleman who was this week on a viſit at Waverley, took the opportunity of examining ſome of the holes in the ſand banks with which that diſtrict abounds. As theſe are undoubtedly bored by bank-martins, and are the places where they avowedly breed, he was in hopes they

F 2   might

might have flept there alfo, and that he
might have furprifed them juft as they
were awaking from their winter flumbers.
When he had dug for fome time, he found
the holes were horizontal, and ferpentine, as
I had obferved before; and that the nefts
were depofited at the inner end, and had
been occupied by broods in former fummers:
but no torpid birds were to be found. He
opened and examined about a dozen holes.
Another gentleman made the fame fearch
many years ago, with as little fuccefs.

Thefe holes were in depth about two
feet.

March 21, 1790. A fingle *bank* or *fand
martin* was feen hovering and playing round
the fand pit at Short heath, where in the
fummer they abound.

April 9, 1793. A fober hind affures us,
that this day, on Wifh-hanger common be-

tween

tween Hedleigh and Frinſham, he ſaw ſeveral *bank-martins* playing in and out, and
hanging before ſome neſt holes in a ſand
hill, where theſe birds uſually neſtle.

This incident confirms my ſuſpicions, that
this ſpecies of hirundo is to be ſeen firſt of
any; and gives great reaſon to ſuppoſe that
they do not leave their wild haunts at all,
but are ſecreted amidſt the clefts and caverns
of thoſe abrupt cliffs where they uſually ſpend
their ſummers.

The late ſevere weather conſidered, it is
not very probable that theſe birds ſhould
have migrated ſo early from a tropical re-
gion, through all theſe cutting winds, and
pinching froſts: but it is eaſy to ſuppoſe that
they may, like bats and flies, have been
awakened by the influence of the ſun, amidſt
their ſecret latebræ where they have ſpent the

F 3   uncomfortable,

uncomfortable, foodlefs months in a torpid ftate, and the profoundeft of flumbers.

There is a large pond at Wifh-hanger which induces thefe *fand-martins* to frequent that diftrict.   For I have ever remarked that they haunt near great waters, either rivers, or lakes.

SWALLOWS, CONGREGATING, AND DISAPPEARANCE OF.

DURING the fevere winds that often prevail late in the fpring, it is not eafy to fay how the hirundines fubfift: for they withdraw themfelves, and are hardly ever feen, nor do any infects appear for their fupport. That they can retire to reft, and fleep away thefe uncomfortable periods, as the bats do, is a matter rather to be fufpected than proved: or do they not rather fpend their
time

time in deep and sheltered vales near waters, where insects are more likely to be found? Certain it is, that hardly any individuals of this genus have at such times been seen for several days together.

September 13, 1791. The congregating flocks of hirundines on the church and tower are very beautiful and amusing! When they fly off all together from the roof, on any alarm, they quite swarm in the air. But they soon settle in heaps, and preening their feathers, and lifting up their wings to admit the sun, seem highly to enjoy the warm situation. Thus they spend the heat of the day, preparing for their emigration, and as it were consulting when and where they are to go. The flight about the church seems to consist chiefly of house-martins, about 400 in number: but there are other places of

rendezvous

rendezvous about the village frequented at
the fame time.

It is remarkable, that though moft of them
fit on the battlements and roof, yet many
hang or cling for fome time by their claws
againft the furface of the walls, in a manner
not practifed by them at any other time of
their remaining with us.

The fwallows feem to delight more in
holding their affemblies on trees.

Nov. 3, 1789. Two fwallows were feen
this morning at Newton vicarage houfe,
hovering and fettling on the roofs and out-
buildings. None have been obferved at
Selborne fince October 11. It is very remarkable, that after the hirundines have dif-
appeared for fome weeks, a few are occafionally feen again. Sometimes, in the firft
week in November, and that only for one

6   day.

day. Do they not withdraw and flumber in fome hiding place during the interval? for we cannot fuppofe they had migrated to warmer climes, and fo returned again for one day. Is it not more probable that they are awakened from fleep, and like the bats are come forth to collect a little food? Bats appear at all feafons, through the autumn and fpring months, when the thermometer is at 50, becaufe then phalænæ, moths, are ftirring.

Thefe fwallows looked like young ones.

WAGTAILS.

WHILE the cows are feeding in moift low paftures, broods of wagtails, white and grey, run round them, clofe up to their nofes, and under their very bellies, availing themfelves of the flies that fettle on their legs, and probably finding worms and larvæ that are

roufed

roufed by the trampling of their feet. Nature is fuch an œconomift, that the moft in-
congruous animals can avail themfelves of
each other! intereft makes ftrange friend-
fhips.

WRYNECK.

Theſe birds appear on the grafs-plots and
walks; they walk a little as well as hop,
and thruft their bills into the turf, in queft, I
conclude, of ants, which are their food. While
they hold their bills in the grafs, they draw
out their prey with their tongues, which are
fo long as to be coiled round their heads.

GROSBEAK.

Mr. B. fhot a cock grofbeak which he
had obferved to haunt his garden for more
than a fortnight. I began to accufe this
bird of making fad havock among the buds

8   of

of the cherries, goofeberries, and wall fruit,
of all the neighbouring orchards. Upon
opening its crop or craw, no buds were to
be feen; but a mafs of kernels of the ftones
of fruits. Mr. B. obferved that this bird
frequented the fpot where plum trees grow;
and that he had feen it with fomewhat hard
in its mouth, which it broke with difficulty;
thefe were the ftones of damfons. The latin
ornithologifts call this bird *coccothraufies,* i. e.
berry-breaker, becaufe with its large, horny
beak, it cracks and breaks the fhells of ftone
fruits for the fake of the feed or kernel.
Birds of this fort are rarely feen in England,
and only in winter.

OBSERVATIONS ON QUADRUPEDS.

SHEEP.

THE sheep on the downs this winter
(1769) are very ragged, and their coats
much torn; the shepherds say they tear their
fleeces with their own mouths and horns,
and that they are always in that way in mild,
wet winters, being teased and tickled with a
kind of lice.

After ewes and lambs are shorn, there is
great confusion and bleating, neither the
dams nor the young being able to distinguish
one another as before. This embarrassment
seems not so much to arise from the loss of
the

the fleece, which may occafion an alteration in their appearance, as from the defect of that *notus odor*, difcriminating each individual perfonally; which alfo is confounded by the ftrong fcent of the pitch and tar wherewith they are newly marked. For the brute creation recognize each other more from the *fmell* than the *fight*; and in matters of identity and diverfity, appeal much more to their nofes, than to their eyes.

After fheep have been wafhed there is the fame confufion, from the reafon given above.

RABBITS.

RABBITS make incomparably the fineft turf; for they not only bite clofer than larger quadrupeds, but they allow no bents to rife: hence warrens produce much the moft delicate turf for gardens. Sheep never touch the ftalks of graffes.

CAT

CAT AND SQUIRRELS.

A boy has taken three little young fquirrels in their neft or *drey* as it is called in thefe
parts. Thefe fmall creatures he put under
the care of a cat who had lately loft her
kittens, and finds that fhe nurfes and fuckles
them with the fame affiduity and affection
as if they were her own offspring. This
circumftance corroborates my fufpicion, that
the mention of expofed and deferted children
being nurtured by female beafts of prey who
had loft their young, may not be fo impro-
bable an incident as many have fuppofed;
and therefore may be a juftification of thofe
authors, who have gravely mentioned, what
fome have deemed to be a wild and impro-
bable ftory.

So many people went to fee the little
fquirrels fuckled by a cat, that the fofter
mother

mother became jealous of her charge, and in pain for their fafety; and therefore hid them over the ceiling, where one died. This circumftance fhews her affection for thefe foundlings, and that fhe fuppofes the fquirrels to be her own young. Thus hens, when they have hatched ducklings, are equally attached to them as if they were their own chickens.

HORSE.

An old hunting mare, which ran on the common, being taken very ill, ran down into the village, as it were to implore the help of men, and died the night following in the ftreet.

HOUNDS.

The king's ftag-hounds came down to Alton, attended by a huntfman, and fix yeo-

men

men prickers, with horns, to try for the ſtag
that has haunted Hartley wood and its en-
virons for ſo long a time. Many hundreds
of people, horſe and foot, attended the dogs
to ſee the deer unharboured; but though the
huntſman drew Hartley wood, and long coppice, and ſhrubwood, and Temple hangers;
and in their way back, Hartley and Ward-le-
ham hangers, yet no ſtag could be found.

The royal pack, accuſtomed to have the
deer turned out before them, never drew the
coverts with any addreſs and ſpirit, as many
people that were preſent obſerved: and this
remark the event has proved to be a true
one. For as a perſon was lately purſuing a
pheaſant that was wing-broken, in Hartley
wood, he ſtumbled upon the ſtag by acci-
dent, and ran in upon him as he lay con-
cealed amidſt a thick brake of brambles and
buſhes.

OBSERVATIONS

OBSERVATIONS ON INSECTS AND VERMES.

INSECTS IN GENERAL.

THE day and night infects occupy the annuals alternately: the papilios, mufcæ, and apes, are fucceeded at the clofe of the day by phalænæ, earwigs. woodlice, &c. In the dufk of the evening, when beetles begin to buz, partridges begin to call; thefe two circumftances are exactly coincident.

Ivy is the laft flower that fupports the hymenopterous and dipterous infects. On funny days quite on to November they fwarm on trees covered with this plant; and when they difappear, probably retire under

G   the

the fhelter of its leaves, concealing themfelves between its fibres and the trees which
it entwines.

Spiders, woodlice, lepifmæ in cupboards
and among fugar, fome empedes, gnats, flies
of feveral fpecies, fome phalænæ in hedges,
earth-worms, &c. are ftirring at all times
when winters are mild; and are of great
fervice to thofe foft-billed birds that never
leave us.

On every funny day the winter through,
clouds of infects ufually called gnats (I fuppofe tipulæ and empedes) appear fporting
and dancing over the tops of the ever-green
trees in the fhrubbery, and frifking about as
if the bufinefs of generation was ftill going
on. Hence it appears that thefe diptera
(which by their fizes appear to be of different fpecies) are not fubject to a torpid ftate
in the winter, as moft winged infects are.

At

At night, and in frofty weather, and when it rains and blows, they feem to retire into thofe trees. They often are out in a fog.

HUMMING IN THE AIR.

THERE is a natural occurrence to be met with upon the higheft part of our down in hot fummer days, which always amufes me much, without giving me any fatisfaction with refpect to the caufe of it; and that is a loud audible humming of bees in the air, though not one infect is to be feen. This found is to be heard diftinctly the whole common through, from the Money-dells, to Mr. White's avenue gate. Any perfon would fuppofe that a large fwarm of bees was in motion, and playing about over his head. This noife was heard laft week, on June 28th.

G 2   " Refounds

" Refounds the living furface of the ground,
Nor undelightful is the ceafelefs *hum*
To him who mufes ——————— at noon."
" Thick in yon ftream of light a thoufand ways,
Upward and downward, thwarting and convolved,
The quivering nations fport."

Thomfon's Seafons.

CHAFFERS.

COCKCHAFFERS feldom abound oftener than once in three or four years; when they fwarm, they deface the trees and hedges. Whole woods of oaks are ftripped bare by them.

Chaffers are eaten by the turkey, the rook, and the houfe-fparrow.

The fcarabæus folftitialis firft appears about June 26: they are very punctual in their coming out every year. They are a fmall fpecies, about half the fize of the *May chaffer*, and are known in fome parts by the name of the *fern chaffer*.

PTINUS

PTINUS PECTINICORNIS.

THOSE maggots that make worm-holes in tables, chairs, bed-pofts, &c. and deftroy wooden furniture, efpecially where there is any fap, are the larvæ of the Ptinus pectinicornis. This infect, it is probable, depofits its eggs on the furface, and the worms eat their way in.

In their holes they turn into their pupæ ftate, and fo come forth winged in July; eating their way through the valances or curtains of a bed, or any other furniture that happens to obftruct their paffage.

They feem to be moft inclined to breed in beech; hence beech will not make lafting utenfils, or furniture. If their eggs are depofited on the furface, frequent rubbings will preferve wooden furniture.

BLATTA ORIENTALIS. COCKROACH.

A NEIGHBOUR complained to me that her houfe was over-run with a kind of *black beetle*, or, as fhe expreffed herfelf, with a kind of *black-bob*, which fwarmed in her kitchen when they get up in a morning before day break.

Soon after this account, I obferved an unufual infect in one of my dark chimney clofets, and find fince, that in the night they fwarm alfo in my kitchen. On examination, I foon afcertained the fpecies to be the *blatta orientalis* of *Linnæus*, and the *blatta molendinaria* of *Mouffet*. The male is winged; the female is not, but fhows fomewhat like the rudiments of wings, as if in the pupa ftate.

Thefe infects belonged originally to the warmer parts of America, and were con-

veyed

veyed from thence by shipping to the East
Indies; and by means of commerce begin
to prevail in the more northern parts of
Europe, as Ruffia, Sweden, &c. How long
they have abounded in England I cannot
say; but have never obferved them in my
houfe till lately.

They love warmth, and haunt chimney-
clofets, and the backs of ovens. *Poda* fays
that thefe and houfe crickets will not affoci-
ate together; but he is miftaken in that af-
fertion, as *Linnæus* fufpected he was. They
are altogether night infects, *lucifugæ*, never
coming forth till the rooms are dark and
ftill, and efcaping away nimbly at the approach of a candle. Their antennæ are remarkably long, flender, and flexile.

October, 1790. After the fervants are
gone to bed, the kitchen hearth fwarms with
young crickets, and young *blattæ molendinariæ*

of

of all fizes, from the moft minute growth to
their full proportions. They feem to live
in a friendly manner together, and not to
prey the one on the other.

Auguft, 1792. After the deftruction of
many thoufands of *blattæ molendinariæ*, we
find that at intervals a frefh detachment
of old ones arrives; and particularly during
this hot feafon : for the windows being left
open in the evenings, the males come flying
in at the cafements from the neighbouring
houfes, which fwarm with them. How the
females, that feem to have no perfect wings
that they can ufe, can contrive to get from
houfe to houfe, does not fo readily appear.
Thefe, like many infects, when they find
their prefent abodes over-ftocked, have
powers of migrating to frefh quarters. Since
the *blattæ* have been fo much kept under,
the crickets have greatly increafed in number.

GRYLLUS

GRYLLUS DOMEST. HOUSE CRICKET.

NOVEMBER. After the fervants are gone to bed, the kitchen hearth fwarms with minute crickets not fo large as fleas, which muft have been lately hatched. So that thefe domeftic infects, cherifhed by the influence of a conftant large fire, regard not the feafon of the year, but produce their young at a time when their congeners are either dead, or laid up for the winter, to pafs away the uncomfortable months in the profoundeft flumbers, and a ftate of torpidity.

When houfe-crickets are out, and running about in a room in the night, if furprifed by a candle, they give two or three fhrill notes, as it were for a fignal to their fellows, that they may efcape to their crannies and lurking holes, to avoid danger.

CIMEX LINEARIS.

AUGUST 12, 1775 *Cimices lineares* are now
in high copulation on ponds and pools. The
females, who vaftly exceed the males in bulk,
dart and fhoot along on the furface of the
water with the males on their backs. When
a female choofes to be difengaged, fhe rears,
and jumps, and plunges, like an unruly colt;
the lover thus difmounted, foon finds a new
mate. The females, as faft as their curiofi-
ties are fatisfied, retire to another part of the
lake, perhaps to depofit their fœtus in quiet;
hence the fexes are found feparate, except
where generation is going on. From the
multitude of minute young of all gradations
of fizes, thefe infects feem without doubt to
be viviparous.

PHALÆNA QUERCUS.

MOST of our oaks are naked of leaves, and even the *Holt* in general, having been ravaged by the caterpillars of a fmall *phalæna* which is of a pale yellow colour. Thefe infects, though a feeble race, yet from their infinite numbers are of wonderful effect, being able to deftroy the foliage of whole forefts and diftricts. At this feafon they leave their *aurelia*, and iffue forth in their *fly*-ftate, fwarming and covering the trees and hedges.

In a field at Greatham, I faw a flight of *fwifts* bufied in catching their prey near the ground ; and found they were hawking after thefe *phalænæ*. The *aurelia* of this moth is fhining and as black as jet; and lies wrapped up in a leaf of the tree, which is rolled round it, and fecured at the ends by a web, to pre vent the maggot from falling out.

EPHEMERA

EPHEMERA CAUDA BISETA. MAY FLY.

JUNE 10, 1771. Myriads of May flies appear for the firſt time on the Alresford ſtream. The air was crowded with them, and the ſurface of the water covered. Large trouts ſucked them in as they lay ſtruggling on the ſurface of the ſtream, unable to riſe till their wings were dried.

This appearance reconciled me in ſome meaſure to the wonderful account that Scopoli gives of the quantities emerging from the rivers of Carniola. Their motions are very peculiar, up and down for many yards almoſt in a perpendicular line.

SPHYNX OCELLATA?

A VAST inſeᴄt appears after it is duſk, flying with a humming noiſe, and inſerting its tongue

tongue into the bloom of the honey-fuckle;
it fcarcely fettles upon the plants, but feeds
on the wing in the manner of humming
birds.

WILD BEE.

THERE is a fort of wild bee frequenting
the garden-campion for the fake of its to-
mentum, which probably it turns to fome
purpofe in the bufinefs of nidification.   It is
very pleafant to fee with what addrefs it
ftrips off the pubes, running from the top to
the bottom of a branch, and fhaving it bare
with all the dexterity of a hoop-fhaver.
When it has got a vaft bundle, almoft as
large as itfelf, it flies away, holding it fecure
between its chin and its fore legs.

There is a remarkable hill on the downs
near *Lewes* in *Suffex,* known by the name of
Mount Carburn, which overlooks that town,
and

and affords a moſt engaging proſpect of all
the country round, beſides ſeveral views of
the ſea.   On the very ſummit of this exalted
promontory, and amidſt the trenches of its
Daniſh camp, there haunts a ſpecies of wild
bee, making its neſt in the chalky ſoil.
When people approach the place, theſe in-
ſects begin to be alarmed, and with a ſharp
and hoſtile ſound, daſh and ſtrike round
the heads and faces of intruders.   I have
often been interrupted myſelf while contem-
plating the grandeur of the ſcenery around
me, and have thought myſelf in danger of
being ſtung.

WASPS.

Wasps abound in woody wild diſtricts
far from neighbourhoods; they feed on
flowers, and catch flies and caterpillars, to
carry to their young. Waſps make their neſts
with

with the rafpings of found timber; hornets,
with what they gnaw from decayed: thefe par-
ticles of wood are kneaded up with a mix-
ture of faliva from their bodies, and moulded
into combs.

When there is no fruit in the gardens,
wafps eat flies, and fuck the honey from
flowers, from ivy bloffoms, and umbellated
plants: they carry off alfo flefh from butchers
fhambles.

OESTRUS CURVICAUDA.

THIS infect lays its nits or eggs on
horfes' legs, flanks, &c. each on a fingle
hair. The maggots when hatched do not
enter the horfes' fkins, but fall to the ground.
It feems to abound moft in moift, moorifh
places, though fometimes feen in the up-
lands.

NOSE

NOSE FLY.

About the beginning of July, a species of fly (musca) obtains, which proves very tormenting to horses, trying still to enter their nostrils and ears, and actually laying their eggs in the latter of those organs, or perhaps in both. When these abound, horses in woodland districts become very impatient at their work, continually tossing their heads, and rubbing their noses on each other, regardless of the driver, so that accidents often ensue. In the heat of the day, men are often obliged to desist from plowing. Saddle-horses are also very troublesome at such seasons. Country people call this insect the *nose fly*.

ICHNEUMON FLY.

I saw lately a small Ichneumon fly attack a spider much larger than itself on a

grass

grafs walk.   When the fpider made any refiftance, the Ichneumon applied her tail to
him, and ftung him with great vehemence,
fo that he foon became dead and motionlefs.
The Ichneumon then running backward drew
her prey very nimbly over the walk into the
ftanding grafs.   This fpider would be depofited in fome hole where the Ichneumon
would lay fome eggs; and as foon as the
eggs were hatched, the carcafe would afford
ready food for the maggots.

Perhaps fome eggs might be injected into
the body of the fpider, in the act of ftinging.
Some Ichneumons depofit their eggs in the
aurelia of moths and butterflies.

BOMBYLIUS MEDIUS.

THE *bombylius medius* is much about in
March and the beginning of April, and foon
feems to retire.   It is an hairy infect, like

H

an

an humble-bee, but with only two wings,
and a long ſtraight beak, with which it
ſucks the early flowers. The female ſeems
to lay its eggs as it poiſes on its wings, by
ſtriking its tail on the ground, and againſt
the graſs that ſtands in its way, in a quick
manner, for ſeveral times together.

MUSCÆ.—FLIES.

In the decline of the year, when the
mornings and evenings become chilly, many
ſpecies of flies (muſcæ) retire into houſes,
and ſwarm in the windows.

At firſt they are very briſk and alert; but
as they grow more torpid, one cannot help
obſerving that they move with difficulty, and
are ſcarce able to lift their legs, which ſeem
as if glued to the glaſs; and by degrees
many do actually ſtick on till they die in the
place.

It

It has been obferved that divers flies, be-
fides their fharp, hooked nails, have alfo
fkinny palms, or flaps to their feet, whereby
they are enabled to ftick on glafs and other
fmooth bodies, and to walk on ceilings with
their backs downward by means of the preffure
of the atmofphere on thofe flaps : the weight
of which they eafily overcome in warm weather when they are brifk and alert. But in the
decline of the year, this refiftance becomes
too mighty for their diminifhed ftrength; and
we fee flies labouring along, and lugging
their feet in windows as if they ftuck faft to
the glafs, and it is with the utmoft difficulty
they can draw one foot after another, and
difengage their hollow caps from the flippery
furface.

Upon the fame principle that flies ftick
and fupport themfelves, do boys, by way of
play, carry heavy weights by only a piece of

wet leather at the end of a string clapped
close on the surface of a stone.

TIPULÆ, or EMPEDES.

MILLIONS of *empedes*, or *tipulæ*, come forth
at the close of day, and swarm to such a
degree as to fill the air. At this juncture
they sport and copulate; as it grows more
dark they retire. All day they hide in the
hedges. As they rise in a cloud they appear
like smoke.

I do not ever remember to have seen such
swarms, except in the fens of the Isle of Ely
They appear most over grass grounds.

ANTS.

AUGUST 23. Every ant-hill about this
time is in a strange hurry and confusion; and
all the winged ants, agitated by some violent
impulse, are leaving their homes, and, bent

on

on emigration, fwarm by myriads in the air, to the great emolument of the hirundines, which fare luxurioufly. Thofe that efcape the fwallows return no more to their nefts, but looking out for frefh fettlements, lay a foundation for future colonies. All the females at this time are pregnant : the males that efcape being eaten, wander away and die.

October 2. Flying-ants, male and female, ufually fwarm and migrate on hot funny days in Auguft and September; but this day a vaft emigration took place in my garden, and myriads came forth, in appearance from the drain which goes under the fruit wall; filling the air and the adjoining trees and fhrubs with their numbers. The females were full of eggs. This late fwarming is probably owing to the backward, wet

H 3   feafon.

feafon. The day following, not one flying
ant was to be feen.

Horfe-ants travel home to their nefts
laden with flies, which they have caught,
and the aureliæ of fmaller ants, which they
feize by violence.

GLOW WORMS.

By obferving two glow-worms which
were brought from the field to the bank in
the garden, it appeared to us, that thefe little
creatures put out their lamps between eleven
and twelve, and fhine no more for the reft of
the night.

Male glow-worms, attracted by the light
of the candles, come into the parlour.

EARTH WORMS.

EARTH worms make their cafts moft in
mild weather about March and April; they

do

do not lie torpid in winter, but come forth
when there is no froft; they travel about in
rainy nights, as appears from their finuous
tracks on the foft muddy foil, perhaps in
fearch of food.

When earth-worms lie out a-nights on the
turf, though they extend their bodies a great
way, they do not quite leave their holes, but
keep the ends of their tails fixed therein, fo
that on the leaft alarm they can retire with
precipitation under the earth. Whatever
food falls within their reach when thus ex-
tended, they feem to be content with, fuch
as blades of grafs, ftraws, fallen leaves, the
ends of which they often draw into their
holes; even in copulation their hinder parts
never quit their holes; fo that no two, except they lie within reach of each other's
bodies, can have any commerce of that kind;
but as every individual is an hermaphrodite,

there

there is no difficulty in meeting with a mate, as would be the cafe were they of different fexes.

SNAILS AND SLUGS.

THE fhell-lefs fnails called flugs are in motion all the winter in mild weather, and commit great depredations on garden plants, and much injure the green wheat, the lofs of which is imputed to earth worms; while the fhelled fnail, the Φερεοικος, does not come forth at all till about April 10th, and not only lays itfelf up pretty early in autumn, in places fecure from froft, but alfo throws out round the mouth of its fhell a thick operculum formed from its own faliva; fo that it is perfectly fecured, and corked up as it were, from all inclemencies. The caufe why the flugs are able to endure the cold fo much better than fhell fnails is,

that

that their bodies are covered with flime as whales are with blubber.

Snails copulate about Midfummer; and foon after, depofit their eggs in the mould by running their heads and bodies under ground. Hence the way to be rid of them is to kill as many as poffible before they begin to breed.

Large, grey, fhell-lefs, cellar fnails lay themfelves up about the fame time with thofe that live abroad; hence it is plain that a defect of warmth is not the only caufe that influences their retreat.

SNAKES SLOUGH.

———There the fnake throws her enamel'd fkin.
Shakefpear, Midf. Night's Dream.

ABOUT the middle of this month (September) we found in a field near a hedge the flough of a large fnake, which feemed

to

to have been newly caſt. From circumſtances it appeared as if turned wrong ſide
outward, and as drawn off backward, like a
ſtocking or woman's glove. Not only the
whole ſkin, but ſcales from the very eyes,
are peeled off, and appear in the head of the
ſlough like a pair of ſpectacles. The reptile, at the time of changing his coat, had
entangled himſelf intricately in the graſs and
weeds, ſo that the friction of the ſtalks and
blades might promote this curious ſhifting
of his exuviæ.

——————"Lubrica ſerpens
"Exuit in ſpinis veſtem." *Lucret.*

It would be a moſt entertaining ſight
could a perſon be an eye-witneſs to ſuch a
feat, and ſee the ſnake in the act of changing his garment. As the convexity of the
ſcales of the eyes in the ſlough is now inward, that circumſtance alone is a proof that

7   the

the fkin has been turned: not to mention
that now the prefent infide is much darker
than the outer. If you look through the
fcales of the fnake's eyes from the concave
fide, viz. as the reptile ufed them, they leffen
objects much.   Thus it appears from what
has been faid, that fnakes crawl out of the
mouth of their own floughs, and quit the
tail part laft, juft as eels are fkinned by a
cook maid.   While the fcales of the eyes
are growing loofe, and a new fkin is forming, the creature, in appearance, muft be
blind, and feel itfelf in an awkward uneafy
fituation.

OBSERVATIONS

OBSERVATIONS ON VEGETABLES.

———————

TREES, ORDER OF LOSING THEIR LEAVES.

ONE of the firſt trees that becomes naked is the walnut; the mulberry, the aſh, eſpecially if it bears many keys, and the horſecheſtnut, come next. All lopped trees, while their heads are young, carry their leaves a long while. Apple-trees and peaches remain green till very late, often till the end of November: young beeches never caſt their leaves till ſpring, till the new leaves ſprout and puſh them off: in the autumn the beechen-leaves turn of a deep cheſnut colour.

colour. Tall beeches caſt their leaves about
the end of October.

SIZE and GROWTH.

MR. Marſham of Stratton, near Norwich,
informs me by letter thus: " I became a
planter early; ſo that an oak which I
planted in 1720 is become now, at 1 foot
from the earth, 12 feet 6 inches in circumference, and at 14 feet (the half of the timber
length) is 8 feet 2 inches. So if the bark
was to be meaſured as timber, the tree gives
116½ feet, buyers meaſure. Perhaps you
never heard of a larger oak while the planter was living. I flatter myſelf that I increaſed the growth by waſhing the ſtem,
and digging a circle as far as I ſuppoſed the
roots to extend, and by ſpreading ſaw-duſt,
&c. as related in the Phil. Tranſ. I wiſh
I had begun with beeches (my favourite
trees

trees as well as yours), I might then have
feen very large trees of my own raifing.
But I did not begin with beech till 1741,
and then by feed; fo that my largeft is now
at five feet from the ground, 6 feet 3 inches
in girth, and with its head fpreads a circle
of 20 yards diameter. This tree was alfo
dug round, wafhed, &c." *Stratton*, 24 *July*,
1790.

The circumference of trees planted by
myfelf at 1 foot from the ground (1790).

				feet.	inches.
Oak in	1730	-	-	4	5
Afh	1730	-	-	4	$6\frac{1}{2}$
Great fir	1751	-	-	5	0
Greateft beech	1751	-	-	4	0
Elm	1750	-	-	5	3
Lime	1756	-	-	5	5

The great oak in the Holt, which is
deemed by Mr. Marfham to be the biggeft
in

in this ifland, at 7 feet from the ground,
meafures in circumference 34 feet. It has
in old times loft feveral of its boughs, and
is tending to decay. Mr. Marfham com-
putes, that at 14 feet length this oak contains 1000 feet of timber.

It has been the received opinion that trees
grow in height only by their annual upper
fhoot. But my neighbour over the way,
whofe occupation confines him to one fpot,
affures me, that trees are expanded and
raifed in the lower parts alfo. The reafon
that he gives is this; the point of one of
my firs began for the firft time to peep over
an oppofite roof at the beginning of fummer; but before the growing feafon was
over, the whole fhoot of the year, and three
or four joints of the body befide, became
vifible to him as he fits on his form in his
fhop. According to this fuppofition, a tree
may

may advance in height confiderably, though the fummer fhoot fhould be deftroyed every year.

FLOWING OF SAP.

IF the bough of a vine is cut late in the fpring, juft before the fhoots pufh out, it will bleed confiderably; but after the leaf is out, any part may be taken off without the leaft inconvenience. So oaks may be barked while the leaf is budding; but as foon as they are expanded, the bark will no longer part from the wood, becaufe the fap that lubricates the bark and makes it part, is evaporated off through the leaves.

RENOVATION OF LEAVES.

WHEN oaks are quite ftripped of their leaves by chaffers, they are clothed again foon after Midfummer with a beautiful foliage :

foliage: but beeches, horfe-cheftnuts and maples, once defaced by thofe infects, never recover their beauty again for the whole feafon.

ASH TREES.

MANY afh trees bear loads of keys every year, others never feem to bear any at all. The prolific ones are naked of leaves and unfightly; thofe that are fteril abound in foliage, and carry their verdure a long while, and are pleafing objects.

SYCAMORE.

MAY 12. The fycamore or great maple is in bloom, and at this feafon makes a beautiful appearance, and affords much pabulum for bees, fmelling ftrongly like honey. The foliage of this tree is very fine, and

I   very

very ornamental to outlets. All the maples have faccharine juices.

GALLS OF LOMBARDY POPLAR.

THE ftalks and ribs of the leaves of the Lombardy poplar are emboffed with large tumours of an oblong fhape, which by in-curious obfervers have been taken for the fruit of the tree. Thefe galls are full of fmall infects, fome of which are winged, and fome not. The parent infect is of the ge-nus of *cynips*. Some poplars in the garden are quite loaded with thefe excrefcences.

CHESTNUT TIMBER.

JOHN CARPENTER brings home fome old cheftnut trees which are very long; in fe-veral places the wood-peckers had begun to bore them. The timber and bark of thefe

7

trees are fo very like oak, as might eafily deceive an indifferent obferver, but the wood is very fhakey, and towards the heart *cup-fhakey*, (that is to fay, apt to feparate in round pieces like cups) fo that the inward parts are of no ufe. They were bought for the purpofe of cooperage, but muft make but ordinary barrels, buckets, &c. Cheftnut fells for half the price of oak; but has fome-times been fent into the king's docks, and paffed off inftead of oak.

LIME BLOSSOMS.

DR. CHANDLER tells that in the fouth of France, an infufion of the bloffoms of the lime tree, *tilia*, is in much efteem as a re-medy for coughs, hoarfneffes, fevers, &c. and that at Nifmes he faw an avenue of limes that was quite ravaged and torn in pieces by people greedily gathering the

I 2   bloom,

bloom, which they dried and kept for thefe purpofes.

Upon the ftrength of this information we made fome tea of lime bloffoms, and found it a very foft, well-flavoured, pleafant, faccharine julep, in tafte much refembling the juice of liquorice.

BLACKTHORN.

THIS tree ufually bloffoms while cold N. E. winds blow; fo that the harfh rugged weather obtaining at this feafon, is called by the country people, blackthorn winter.

IVY BERRIES.

Ivy berries afford a noble and providential fupply for birds in winter and fpring! for the firft fevere froft freezes and fpoils all the haws, fometimes by the middle of November; ivy berries do not feem to freeze.

HOPS.

HOPS.

THE culture of Virgil's vines correfponded very exactly with the modern management of hops. I might inftance in the perpetual diggings and hoeings, in the tying to the ftakes, and poles, in pruning the fuperfluous fhoots, &c. but lately I have obferved a new circumftance, which was a neighbouring farmer's harrowing between the rows of hops with a fmall triangular harrow, drawn by one horfe, and guided by two handles. This occurrence brought to my mind the following paffage,

——————————————" ipfa
" Flectere luctantes inter vineta juvencos."
Georgic. II.

Hops are diécious plants: hence perhaps it might be proper, though not practifed, to leave purpofely fome male plants in every

I 3   garden,

garden, that their farina might impregnate the bloſſoms. The female plants without their male attendants are not in their natural ſtate: hence we may ſuppoſe the frequent failure of crop ſo incident to hop-grounds; no other growth, cultivated by man, has ſuch frequent and general failures as hops.

Two hop gardens much injured by a hail ſtorm, June 5, ſhew now (September 2) a prodigious crop, and larger and fairer hops than any in the pariſh. The owners ſeem now to be convinced that the hail, by beat-ing off the tops of the binds, has increaſed the ſide-ſhoots, and improved the crop. Query. Therefore ſhould not the tops of hops be pinched off when the binds are very groſs, and ſtrong?

SEED LYING DORMANT.

THE naked part of the Hanger is now co-vered with thiſtles of various kinds. The

ſeeds

feeds of thefe thiftles may have lain probably under the thick fhade of the beeches
for many years, but could not vegetate till
the fun and air were admitted. When old
beech trees are cleared away, the naked
ground in a year or two becomes covered
with ftrawberry plants, the feeds of which
muft have lain in the ground for an age at
leaft. One of the *flidders* or trenches down
the middle of the Hanger, clofe covered over
with lofty beeches near a century old, is ftill
called *ftrawberry flidder*, though no ftraw-
berries have grown there in the memory of
man. That fort of fruit did once, no doubt,
abound there, and will again when the ob-
ftruction is removed.

BEANS SOWN BY BIRDS.

MANY horfe-beans fprang up in my field-
walks in the autumn, and are now grown to

I 4   a confiderable

a confiderable height. As the Ewel was in beans laft fummer, it is moft likely that thefe feeds came from thence; but then the diftance is too confiderable for them to have been conveyed by mice. It is moft probable therefore that they were brought by birds, and in particular by jays and pies, who feem to have hid them among the grafs and mofs, and then to have forgotten where they had ftowed them. Some peafe are growing alfo in the fame fituation, and probly under the fame circumftances.

CUCUMBERS SET BY BEES.

IF bees, who are much the beft fetters of cucumbers, do not happen to take kindly to the frames, the beft way is to tempt them by a little honey put on the male and female bloom. When they are once induced to haunt the frames, they fet all the fruit, and

will

will hover with impatience round the lights in a morning, till the glaffes are opened. Probatum eft.

WHEAT.

A NOTION has always obtained, that in England hot fummers are productive of fine crops of wheat; yet in the years 1780 and 1781, though the heat was intenfe, the wheat was much mildewed, and the crop light. Does not fevere heat, while the ftraw is milky, occafion its juices to exfude, which being extravafated, occafion fpots, difcolour the ftems and blades, and injure the health of the plants?

TRUFFLES.

AUGUST. A truffle-hunter called on us, having in his pocket feveral large truffles found in this neighbourhood. He fays thefe

roots

roots are not to be found in deep woods, but in narrow hedge-rows and the fkirts of coppices. Some truffles, he informed us, lie two feet within the earth, and fome quite on the furface; the latter, he added, have little or no fmell, and are not fo eafily difcovered by the dogs as thofe that lie deeper. Half a crown a pound was the price which he afked for this commodity.

Truffles never abound in wet winters and fprings. They are in feafon in different fituations, at leaft nine months in the year.

TREMELLA NOSTOC.

Though the weather may have been ever fo dry and burning, yet after two or three wet days, this jelly-like fubftance abounds on the walks.

FAIRY

FAIRY RINGS.

THE caufe, occafion, call it what you will, of *fairy-rings*, fubfifts in the turf, and is conveyable with it: for the turf of my garden-walks, brought from the down above, abounds with thofe appearances, which vary their fhape, and fhift fituation continually, difcovering themfelves now in circles, now in fegments, and fometimes in irregular patches and fpots. Wherever they obtain, puff-balls abound; the feeds of which were doubtlefs brought in the turf.

METEOROLOGICAL OBSERVATIONS.

BAROMETER.

NOVEMBER 22, 1768. A remarkable fall of the barometer all over the kingdom. At Selborne we had no wind, and not much rain; only vaft, fwagging, rock-like clouds, appeared at a diftance.

PARTIAL FROST.

THE country people, who are abroad in winter mornings long before fun-rife, talk much of hard froft in fome fpots, and none in others. The reafon of thefe partial frofts is obvious, for there are at fuch times par-

tial

tial fogs about; where the fog obtains, little
or no froft appears: but where the air is
clear, there it freezes hard. So the froft
takes place either on hill or in dale, where-
ever the air happens to be cleareft, and freeft
from vapour.

THAW.

THAWS are fometimes furprifingly quick,
confidering the fmall quantity of rain. Does
not the warmth at fuch times come from below? The cold in ftill, fevere feafons feems
to come down from above: for the coming
over of a cloud in fevere nights raifes the
thermometer abroad at once full ten degrees. The firft notices of thaws often feem
to appear in vaults, cellars, &c.

If a froft happens, even when the ground
is confiderably dry, as foon as a thaw takes
place,

place, the paths and fields are all in a batter. Country people fay that the froft draws moifture. But the true philofophy is, that the fteam and vapours continually afcending from the earth, are bound in by the froft, and not fuffered to efcape till releafed by the thaw. No wonder then that the furface is all in a float; fince the quantity of moifture by evaporation that arifes daily from every acre of ground is aftonifhing.

FROZEN SLEET.

JANUARY 20. Mr. H.'s man fays that he caught this day in a lane near Hackwood park, many rooks, which, attempting to fly, fell from the trees with their wings frozen together by the fleet, that froze as it fell. There were, he affirms, many dozen fo difabled.

MIST,

MIST, CALLED LONDON SMOKE

THIS is a blue mift which has fome-what the fmell of coal fmoke, and as it always comes to us with a N. E. wind, is fuppofed to come from London. It has a ftrong fmell, and is fuppofed to occafion blights. When fuch mifts appear they are ufually followed by dry weather.

REFLECTION OF FOG.

WHEN people walk in a deep white fog by night with a lanthorn, if they will turn their backs to the light, they will fee their fhades impreffed on the fog in rude gigantic pro-portions. This phenomenon feems not to have been attended to, but implies the great denfity of the meteor at that juncture.

HONEY

HONEY DEW.

JUNE 4, 1783. Vaſt honey dews this week. The reaſon of theſe ſeem to be, that in hot days the effluvia of flowers are drawn up by a briſk evaporation, and then in the night fall down with the dews with which they are entangled.

This clammy ſubſtance is very grateful to bees, who gather it with great aſſiduity, but it is injurious to the trees on which it hap-pens to fall, by ſtopping the pores of the leaves. The greateſt quantity falls in ſtill cloſe weather; becauſe winds diſperſe it, and copious dews dilute it, and prevent its ill effects. It falls moſtly in hazy warm wea-ther.

MORNING

MORNING CLOUDS.

AFTER a bright night and vaſt dew, the ſky uſually becomes cloudy by eleven or twelve o'clock in the forenoon, and clear again towards the decline of the day. The reaſon ſeems to be, that the dew, drawn up by evaporation, occaſions the clouds; which, towards evening, being no longer rendered buoyant by the warmth of the ſun, melt away, and fall down again in dews. If clouds are watched in a ſtill warm evening, they will be ſeen to melt away, and diſappear.

DRIPPING WEATHER AFTER DROUGHT.

No one that has not attended to ſuch matters, and taken down remarks, can be aware how much ten days dripping weather

K   will

will influence the growth of grafs or corn
after a fevere dry feafon. This prefent fummer, 1776, yielded a remarkable inftance;
for till the 30th of May the fields were
burnt up and naked, and the barley not half
out of the ground; but now, June 10, there
is an agreeable profpect of plenty.

AURORA BOREALIS.

NOVEMBER 1, 1787. The N. aurora
made a particular appearance, forming itfelf into a broad, red, fiery belt, which ex-
tended from E. to W. acrofs the welkin:
but the moon rifing at about ten o'clock,
in unclouded majefty, in the E. put an end
to this grand, but awful meteorous phenomenon.

BLACK

BLACK SPRING, 1771.

DR. JOHNSON fays, that " in 1771 the feafon was fo fevere in the ifland of Sky, that it is remembered by the name of the *black fpring*. The fnow, which feldom lies at all, covered the ground for eight weeks, many cattle died, and thofe that furvived were fo emaciated that they did not require the male at the ufual feafon." The cafe was juft the fame with us here in the fouth; never were fo many barren cows known as in the fpring following that dreadful period. Whole dairies miffed being in calf together.

At the end of March the face of the earth was naked to a furprifing degree. Wheat hardly to be feen, and no figns of any grafs; turneps all gone, and fheep in a ftarving way. All provifions rifing in price. Farmers cannot fow for want of rain.

K 2

ON THE DARK, STILL, DRY, WARM
WEATHER

OCCASIONALLY HAPPENING IN THE WINTER MONTHS.

Th' imprifon'd winds flumber within their caves
Faft bound: the fickle vane, emblem of change,
Wavers no more, long-fettling to a point.
   All nature nodding feems compos'd: thick fteams
From land, from flood up-drawn, dimming the day,
" Like a dark ceiling ftand :" flow thro' the air
Goffamer floats, or ftretch'd from blade to blade
The wavy net-work whitens all the field.
   Pufh'd by the weightier atmofphere, up fprings
The ponderous Mercury, from fcale to fcale
Mounting, amidft the Torricellian tube *.
   While high in air, and pois'd upon his wings
Unfeen, the foft, enamour'd wood-lark runs
Thro' all his maze of melody ;—the brake
Loud with the black-bird's bolder note refounds.
   Sooth'd by the genial warmth, the cawing rook
Anticipates the fpring, feleds her mate,
Haunts her tall neft-trees, and with fedulous care
Repairs her wicker eyrie, tempeft-torn.

* The Barometer.

The

The plough-man inly fmiles to fee up turn
His mellow glebe, beft pledge of future crop :
With glee the gardener eyes his fmoking beds :
E'en pining ficknefs feels a fhort relief.

The happy fchool-boy brings tranfported forth
His long-forgotten fcourge, and giddy gig :
O'er the white paths he whirls the rolling hoop,
Or triumphs in the dufty fields of taw.

Not fo the mufeful fage :—abroad he walks
Contemplative, if haply he may find
What caufe controls the tempeft's rage, or whence
Amidft the favage feafon winter fmiles.

For days, for weeks prevails the placid calm.
At length fome drops prelude a change : the fun
With ray refracted burfts the parting gloom ;
When all the chequer'd fky is one bright glare.

Mutters the wind at eve : th' horizon round
With angry afpect fcowls : down rufh the fhowers,
And float the delug'd paths, and miry fields.

K 3

SUMMARY

OF THE

WEATHER.

K 4

Meafure of Rain in Inches and Hundreds.

Year.	Jan.	Feb.	March.	April.	May.	June.	July .	Aug.	Sept.	Oct.	Nov.	Dec.	Total.
1782.	4.64	1.98	6.54	4.57	6.34	1.75	7.09	8.28	3.72	1.93	2.51	0.91	50.26
1783.	4.43	5.54	2.16	0.88	2.84	2.82	1.45	2.24	5.52	1.71	3.01	1.10	33.71
1784.	3.18	0.77	3.82	3.92	1.52	3.65	2.40	3.88	2.51	0.39	4.70	3.06	33.80
1785.	2.84	1.80	0.30	0.17	0.60	1.39	3.80	3.21	5.94	5.21	2.27	4. 2	31.55
1786.	6.91	1.42	1.62	1.81	2.40	1.20	1.99	4.34	4.79	5. 4	4.38	—	
1787.	0.88	3.67	4.28	0.74	2.60	1.50	6.53	0.83	1.56	5. 4	4. 9	5. 6	36.24
1788.	1.60	3.37	1.31	0.61	0.76	1.27	3.58	3.22	5.71	0. 0	0.86	0.23	22.50
1789.	4.48	4.11	2.47	1.81	4. 5	4.24	3.69	0.99	2.82	5. 4	3.67	4.62	42.
1790.	1.99	0.49	0.45	3.64	4.38	0.13	3.24	2.30	0.66	2.10	6.95	5.94	32.27
1791.	6.73	4.64	1.59	1.13	1.33	0.91	5.56	1.73	1.73	6.49	8.16	4.93	44.93
1792.	6. 7	1.68	6.70	4.08	3.00	2.78	5.16	4.25	5.53	5.55	1.65	2.11	48.56
1793.	3.71	2.32	3.33	3.19	1.21								

(153)

SUMMARY

OF THE

WEATHER.

1768 begins with a fortnight's froſt and ſnow; rainy during February. Cold and wet ſpring; wet ſeaſon from the beginning of June to the end of harveſt. Latter end of September foggy, without rain. All October and the firſt part of November rainy; and thence to the end of the year alternate rains and froſts.

1769. January and February, froſty and rainy, with gleams of fine weather in the intervals. To the middle of March, wind and rain. To the end of March, dry and windy. To the middle of April, ſtormy,

with

with rain. To the end of June, fine weather, with rain. To the beginning of Auguſt, warm, dry weather. To the end of September, rainy with ſhort intervals of fine weather. To the latter end of October, froſty mornings, with fine days. The next fortnight rainy; thence to the end of November dry and froſty. December, windy, with rain and intervals of froſt, and the firſt fortnight very foggy.

1770. Froſt for the firſt fortnight: during the 14th and 15th all the ſnow melted. To the end of February, mild hazy weather. The whole of March froſty, with bright weather. April, cloudy, with rain and ſnow. May began with ſummer ſhowers, and ended with dark cold rains. June, rainy, checquered with gleams of ſunſhine. The firſt fortnight in July, dark and ſultry; the latter part of the month, heavy rain.

Auguſt,

Auguft, September, and the firft fortnight in October, in general fine weather, though with frequent interruptions of rain: from the middle of October to the end of the year almoft inceffant rains.

1771. Severe froft till the laft week in January. To the firft week in February, rain and fnow: to the end of February, fpring weather. To the end of the third week in April, frofty weather. To the end of the firft fortnight in May, fpring weather, with copious fhowers. To the end of June, dry, warm weather. The firft fortnight in July, warm, rainy weather. To the end of September, warm weather, but in general cloudy, with fhowers. October, rainy. November, froft, with intervals of fog and rain. December, in general bright, mild weather, with hoar frofts.

1772. To the end of the firft week in
February,

February, froft and fnow. To the end of the firft fortnight in March, froft, fleet, rain and fnow. To the middle of April, cold rains. To the middle of May, dry weather, with cold piercing winds. To the end of the firft week in June, cool fhowers. To the middle of Auguft, hot dry fummer weather. To the end of September, rain with ftorms and thunder. To December 22, rain with mild weather. December 23, the firft ice. To the end of the month, cold foggy weather.

1773. The firft week in January, froft; thence to the end of the month, dark rainy weather. The firft fortnight in February, hard froft. To the end of the firft week in March, mifty, fhowery weather. Bright fpring days to the clofe of the month. Frequent fhowers to the latter end of April. To the end of June, warm fhowers, with

7   intervals

intervals of funfhine. To the end of Auguft, dry weather, with a few days of rain. To the end of the firft fortnight in November, rainy. The next four weeks, froft: and thence to the end of the year, rainy.

1774. Froft and rain to the end of the firft fortnight in March: thence to the end of the month, dry weather. To the 15th of April, fhowers; thence to the end of April, fine fpring days. During May, fhowers and funfhine in about an equal proportion. Dark rainy weather to the end of the 3d week in July: thence to the 24th of Auguft, fultry, with thunder and occafional fhowers. To the end of the 3d week in November, rain, with frequent intervals of funny weather. To the end of December, dark dripping fogs.

1775. To the end of the firft fortnight in March, rain almoft every day. To the

firft

firſt week in April, cold winds, with ſhowers of rain and ſnow. To the end of June, warm, bright weather, with frequent ſhowers. The firſt fortnight in July, almoſt inceſſant rains. To the 26th Auguſt, ſultry weather with frequent ſhowers. To the end of the 3d week in September, rain, with a few intervals of fine weather. To the end of the year, rain, with intervals of hoar-froſt and ſunſhine.

1776. To January 24, dark froſty weather, with much ſnow. March 24, to the end of the month, foggy, with hoar froſt. To the 30th of May, dark, dry harſh weather, with cold winds. To the end of the firſt fortnight in July, warm, with much rain. To the end of the firſt week in Auguſt, hot and dry, with intervals of thunder ſhowers. To the end of October, in general fine ſeaſonable weather, with a conſider

able

able proportion of rain. To the end of the year, dry, frosty weather, with some days of hard rain.

1777. To the 10th of January, hard frost. To the 20th of January, foggy, with frequent showers. To the 18th of February, hard dry frost with snow. To the end of May, heavy showers, with intervals of warm dry spring days. To the 8th July, dark, with heavy rain. To the 18th July, dry, warm weather. To the end of July, very heavy rains. To the 12th October, remarkably fine warm weather. To the end of the year, grey mild weather, with but little rain, and still less frost.

1778. To the 13th of January, frost, with a little snow: to the 24th January, rain: to the 30th, hard frost. To the 23d February, dark, harsh, foggy weather, with rain. To the end of the month, hard frost,

with

with ſnow.   To the end of the firſt fortnight
in March, dark, harſh weather.   From the
firſt, to the end of the firſt fortnight in
April, ſpring weather.   To the end of the
month, ſnow and ice.   To the 11th of
June, cool, with heavy ſhowers.   To the
19th July, hot, ſultry, parching weather.
To the end of the month, heavy ſhowers.
To the end of September, dry warm weather.
To end of the year, wet, with conſiderable
intervals of ſunſhine.

1779.   Froſt and ſhowers to the end of
January.   To 21ſt April, warm dry weather.   To 8th May, rainy.   To the 7th
June, dry and warm.   To the 6th July, hot
weather, with frequent rain.   To the 18th
July, dry hot weather.   To Auguſt 8, hot
weather, with frequent rains.   To the end
of Auguſt, fine dry harveſt weather.   To
the end of November, fine autumnal weather,
                                             with

with intervals of rain. To the end of the year, rain with froſt and ſnow.

1780. To the end of January, froſt. To the end of February, dark, harſh weather, with frequent intervals of froſt. To the end of March, warm ſhowery ſpring weather. To the end of April, dark harſh weather, with rain and froſt. To the end of the firſt fortnight in May, mild, with rain. To the end of Auguſt, rain and fair weather in pretty equal proportions. To the end of October, fine autumnal weather, with intervals of rain. To the 24th November froſt. To December 16, mild dry foggy weather. To the end of the year froſt and ſnow.

1781. To January 25, froſt and ſnow. To the end of February, harſh and windy with rain and ſnow. To April 5, cold drying winds. To the end of May, mild ſpring weather, with a few light ſhowers.

L   June

June began with heavy rain, but thence to the end of October, dry weather, with a few flying showers. To the end of the year, open weather with frequent rains.

1782. To February 4, open mild weather. To February 22, hard frost. To the end of March, cold blowing weather, with frost and snow and rain. To May 7, cold dark rains. To the end of May, mild, with incessant rains. To the end of June, warm and dry. To the end of August warm, with almost perpetual rains. The first fortnight in September mild and dry; thence to the end of the month, rain. To the end of October, mild with frequent showers. November began with hard frost, and continued throughout with alternate frost and thaw. The first part of December frosty; the latter part mild.

1783. To January 16, rainy with heavy winds.

winds. To the 24th, hard froft. To the end of the firft fortnight in February, blowing, with much rain. To the end of February, ftormy dripping weather. To the 9th of May, cold harfh winds (thick ice on 5th of May). To the end of Auguft, hot weather, with frequent fhowers. To the 23d September, mild, with heavy driving rains. To November 12, dry, mild weather. To the 18th December, grey foft weather, with a few fhowers. To the end of the year, hard froft.

1784. To February 19, hard froft, with two thaws; one the 14th January, the other 5th February. To February 28, mild wet fogs. To the 3d March, froft with ice. To March 10, fleet and fnow. To April 2, fnow and hard froft. To April 27, mild weather with much rain. To May 12, cold drying winds. To May 20, hot cloud-

L 2

lefs

lefs weather. To June 27, warm with frequent fhowers. To July 18, hot and dry. To the end of Auguft, warm with heavy rains. To November 6, clear mild autumnal weather, except a few days of rain at the latter end of September. To the end of the year, fog, rain, and hard froft (on December 10, the therm. 1 deg. below 0).

1785. A thaw began on the 2d January, and rainy weather with wind continued to January 28. To 15th March, very hard froft. To 21ft March, mild with fprinkling fhowers. To April 7, hard froft. To May 17, mild windy weather, without a drop of rain. To the end of May, cold with a few fhowers. To June 9, mild weather, with frequent foft fhowers. To July 13, hot dry weather, with a few fhowery intervals. To July 22, heavy rain. To the end of September, warm with frequent

quent showers. To the end of October, frequent rain. To 18th of November, dry, mild weather. (Hay-making finished November 9, and the wheat harvest November 14.) To December 23, rain. To the end of the year, hard frost.

1786. To the 7th January, frost and snow. To January 13, mild with much rain. To 21st January, deep snow. To February 11, mild with frequent rains. To 21st February, dry, with high winds. To 10th March, hard frost. To 13th April, wet, with intervals of frost. To the end of April, dry mild weather. On the 1st and 2d May, thick ice. To 10th May, heavy rain. To June 14, fine warm dry weather. From the 8th to the 11th July, heavy showers. To October 13, warm, with frequent showers. To October 19, ice. To October 24, mild pleasant wea-

L 3   ther.

ther. To November 3, froft. To December 16, rain, with a few detached days of froft. To the end of the year, froft and fnow.

1787. To January 24, dark, moift, mild weather. To January 28, froft and fnow. To February 16, mild fhowery weather. To February 28, dry, cool weather. To March 10, ftormy, with driving rain. To March 24, bright frofty weather. To the end of April, mild, with frequent rain. To May 22, fine bright weather. To the end of June, moftly warm, with frequent fhowers (on June 7, ice as thick as a crown piece). To the end of July, hot and fultry, with copious rain. To the end of September, hot dry weather, with occafional fhowers. To November 23, mild, with light frofts and rain. To the end of November, hard froft. To De-

8   cember

cember 21, ſtill and mild, with rain. To
the end of the year, froſt.

1788. To January 13, mild and wet.
To January 18, froſt. To the end of the
month, dry windy weather. To the end
of February, froſty, with frequent ſhowers.
To March 14, hard froſt. To the end of
March, dark, harſh weather, with frequent
ſhowers. To April 4, windy, with ſhowers.
To the end of May, bright, dry, warm
weather, with a few occaſional ſhowers.
From June 28 to July 17, heavy rains.
To Auguſt 12, hot dry weather. To the
end of September, alternate ſhowers and
ſunſhine. To November 22, dry cool
weather. To the end of the year, hard
froſt.

1789. To January 13, hard froſt. To
the end of the month, mild, with ſhowers.
To the end of February, frequent rain, with

L 4 ſnow-

fnow-fhowers and heavy gales of wind.   To
13th March, hard froft, with fnow.   To
April 18, heavy rain, with froft and fnow
and fleet.   To the end of April, dark cold
weather, with frequent rains.   To June 9,
warm fpring weather, with brifk winds and
frequent fhowers.   From June 4, to the
end of July, warm, with much rain.   To
Auguft 29, hot, dry, fultry weather.   To
September 11, mild, with frequent fhowers.
To the end of September, fine autumnal
weather, with occafional fhowers.   To November 17, heavy rain, with violent gales
of wind.   To December 18, mild dry weather, with a few fhowers.   To the end of
the year, rain and wind.

1790.   To January 16, mild foggy weather, with occafional rains.   To January
21, froft.   To January 28, dark, with driving
rains.   To February 14, mild, dry weather.

To

To February 22, hard froft. To April 5, bright cold weather, with a few fhowers. To April 15, dark and harfh, with a deep fnow. To April 21, cold cloudy weather, with ice. To June 6, mild fpring weather, with much rain. From July 3, to July 14, cool, with heavy rain. To the end of July, warm, dry weather. To Auguft 6, cold, with wind and rain. To Auguft 24, fine harveft weather. To September. 5, ftrong gales, with driving fhowers. To November 26, mild autumnal weather, with frequent fhowers. To December 1, hard froft and fnow. To the end of the year, rain and fnow, and a few days of froft.

1791. To the end of January, mild, with heavy rains. To the end of February windy, with much rain and fnow. From March to the end of June, moftly dry, efpecially June. March and April rather cold

and

and frofty. May and June, hot. July, rainy. Fine harveft weather, and pretty dry, to the end of September. Wet October, and cold towards the end. Very wet and ftormy in November. Much froft in December.

1792. Some hard froft in January, but moftly wet and mild. February, fome hard froft and a little fnow. March, wet and cold. April, great ftorms on the 13th, then fome very warm weather. May and June, cold and dry. July, wet and cool; indifferent harveft, rather late and wet. September, windy and wet. October, fhowery and mild. November, dry and fine. December, mild.

TABLE

OF

CONTENTS.

————

Fern-

CONTENTS.

Ephemera

CONTENTS,

CONTENTS.

Aurora

C O N T E N T S.

THE END.

Printed in the United States
By Bookmasters